Essential Aerodynamics and its Applications

Essential Aerodynamics and its Applications

Edited by **Russell Mikel**

LANRYE
INTERNATIONAL

New Jersey

Published by Clanrye International,
55 Van Reypen Street,
Jersey City, NJ 07306, USA
www.clanryeinternational.com

Essential Aerodynamics and its Applications
Edited by Russell Mikel

International Standard Book Number: 978-1-63240-222-6 (Hardback)

Contents

Preface

The main aim of this book is to educate learners and enhance their research focus by presenting diverse topics covering this vast field. This is an advanced book which compiles significant studies by distinguished experts in the area of analysis. This book addresses successive solutions to the challenges arising in the area of application, along with it; the book provides scope for future developments.

Aerodynamics is a field of dynamics related to the study of the properties of moving air and the interaction between the air and solid bodies moving through it. From the contemporary point of view, aerodynamics is a branch of physics that studies physical laws and their operations related to the displacement of a body into a fluid. This concept can be applied to any body moving in a fluid which is at rest or vice versa. It discusses various topics related to mobile and immobile aerodynamics, wave formation and distribution, flow check method and sports aerodynamics. This book will be beneficial for teachers and researchers who work in the various branches of this field ranging from experimental methods to computational procedures.

It was a great honour to edit this book, though there were challenges, as it involved a lot of communication and networking between me and the editorial team. However, the end result was this all-inclusive book covering diverse themes in the field.

Finally, it is important to acknowledge the efforts of the contributors for their excellent chapters, through which a wide variety of issues have been addressed. I would also like to thank my colleagues for their valuable feedback during the making of this book.

Editor

Section 1

Sport Aerodynamics

Aerodynamic Design of Sports Garments

Harun Chowdhury
School of Aerospace,
Mechanical and Manufacturing Engineering, RMIT University
Australia

1. Introduction

The suitability of materials for sports garments must meet a range of performance parameters for the specific sport. Therefore, the minimisation of detrimental effects of sports garments on sporting performance is becoming an important aspect of sports technology and product design. The demands on sports garments design for high performance sports and particularly those associated with aerodynamic resistance and its associated energy loss during the sporting action extend the conventional design methodology of synthesis, form, and function to new requirements for quantitative understandings of materials performance and fabric construction. Therefore, sports garments design and its engineering modelling require detailed information on the fabric and surface physics of the materials about the aerodynamic behaviour.

A number of factors identified by prior studies (Grappe et al., 2005; Lukes et al., 2005; Chowdhury et al., 2010) that may contribute to the aerodynamic efficiency of athletes in higher speed sports. Some major factors are:

- Athlete body position during activity
- Sport equipment and accessories used in sporting event
- Sports garments

There are several factors that can affect the aerodynamic characteristics of sports garments. These are:

- Speed
- Body position
- Fabric properties
- Garment construction
- Fitting of the garments

Speed is an important parameter for the aerodynamic optimisation of athlete's performance in sports. It has significant influences on the overall aerodynamic efficiency based on the body position and geometric shapes of sports equipment including sports garments. For example, speed has significant effect on aerodynamic properties as the airflow regime may notably change from laminar to turbulent flow at a particular speed. Table 1 listed the average speed in air for various higher speed sports.

Type of sports	Average speed range (km/h)
Ski Jumping	90-100
Cycling	30-60
Downhill Skiing	80 – 120
Speed skating	50
Sprint (running)	32
Swimming	112 (i.e. 2.1 m/s in water)

Table 1. Average speeds in air for various higher speed sports

In all higher speed sports, the body position also plays an important role as some body parts are responsible for generating either pure aerodynamic drag or both drag and lift simultaneously. The angle of attack (inclination angle) on the aerodynamic drag and lift generation can play a crucial role. Generally, the majority of the aerodynamic drag generated by the body shapes is predominantly form or pressure drag. Therefore, the projected frontal area can be a significant factor in aerodynamic drag generation. Grappe et al. (1997) pointed out that the position of the athlete body should be constantly monitored and analysed in order to ensure that the athlete adopts the most aerodynamic position alone with the equipment at all times during the event.

The cyclists and ski jumpers body positions have been extensively studied and analysed by the bio-mechanists since long. Kyle et al. (2004) and Di Prampero et al. (1979) demonstrated that some positions are more aerodynamically efficient than others. Capelli et al. (1993) showed that the importance of the drag reduction for the bicycle itself. During the sporting event, different body part experiences the incoming air differently. In ski jumping, the jumpers try to streamline their body to minimise the frontal area as well as the aerodynamic drag during in-run phase. During flight phase, the trunk, arms and legs are positioned in such a way that the jumper can jump a longer distance. Thus, the minimisation of drag and maximisation of lift can be done depending on the body position during the flight. In cycling, the cyclist body parts face on coming wind differently (e.g., various angles of attack). Therefore, the aerodynamic drag reduction should be based on the posture of body parts and streamlining the body parts as practical as possible. Under the optimal body configuration in higher speed sports, the additional aerodynamic advantages (e.g., reduction of drags) can be possible by manipulating the flow regimes over the body parts using appropriate fabric.

Sports garments are generally made of knitted fabrics that are manufactured with various fabric manufacturing techniques with natural or synthetic materials (e.g., cotton, polymer). The parameters affecting the aerodynamic properties (drag and lift) of fabrics are the surface roughness, seam position, wale orientation and the air permeability (Chowdhury et al., 2009, 2010; Oggiano et al., 2004; Moria et al., 2010). Studies by Dias & Delkumburewatte (2007) reported that there is a relationship between the surface profile and fabric construction parameters. The surface roughness and air permeability of the fabric can be controlled by manipulating different knitting parameters (Oggiano et al., 2009; Spencer, 2001). Hoerner (1952) reported that aerodynamic drag and lift might have direct relation with the air permeability of the fabrics.

The surface profile and the air permeability of sports fabrics can potentially exhibit significant influences on aerodynamic properties (lift and drag) and flow transitions from laminar to turbulent. Surface roughness is an important parameter for lift and drag due to the transitional properties at the boundary layer. Sport fabrics represent a wide spectrum of surface topologies and wide boundary layer behaviours. The importance of the aerodynamic attributes of the fabric materials used in the garment manufacturing was highlighted in numerous studies (Kyle et al., 2004; Brownlie, 1992; Oggiano et al., 2004; Moria et al., 2010). Also fabric surface is arrayed with regular pattern of stitches. The orientation of the stitches or wales within the garment can potentially have effects in the aerodynamic properties. No study has been reported addressing these parameters of garments in the literature.

Sports garments are made of multiple pieces or panels of fabric that are joined together by using seams or fasteners (e.g., zipper, buttons). The prominence of the seam (position and size) might have effect on drag and lift as it may change air flow regime locally. Kyle et al. (2004) mentioned that the reduction of aerodynamic drag of bicycle racing garments is possible by using different fabrics in different zones of the body parts by taking advantage of aerodynamic behaviour. Brownlie (1992) studied the effects of trip wire at transition point for drag reduction. In ski jumping suit, seam is significantly larger in size compared to cycling garment (skin suit) as the thickness of the ski jumping suit material is approximately 10 times larger. The seam in various speed sports garments potentially has significant effect on aerodynamic performance. The effect on the aerodynamic behaviour of the seam on the garments has not been well studied. The seam on sports garments plays a vital role in aerodynamic drag and lift. Therefore, a thorough study on seam should be undertaken in order to utilize its aerodynamic advantages and minimise its negative impact.

Prior studies were primarily carried out in wind tunnels utilizing mainly cylinders and mannequins with athletic garment and fabric covered. Kyle et al. (2004), Brownlie (1992), Moria et al. (2010) and Oggiano et al. (2009) used wind tunnel for drag measurement using vertical cylinder model. But different body parts have different angle configurations. Therefore, it is necessary to evaluate the aerodynamic properties of different parts of the athlete body with various angular configurations depending on the physical characteristics of the sports.

The body configuration of human is extremely complex due to varied shapes and sizes. Although several researchers have studied the aerodynamic characteristics of human bodies, their studies were mainly based on over simplified human body (Hoerner 1965; Brownlie, 1992). Brownlie (1992) used a simplified human analogous model for wind tunnel study of different fabrics. He developed the model with 11 separate components of uniform circular cylinders of different dimensions. Oggiano et al. (2009) tested several fabric samples in the wind tunnel using both the cylinder and leg models. The study showed that the aerodynamic behaviour was similar with the cylinder and the leg models. Chowdhury et al. (2010) demonstrated that athlete body parts can be represented as multiple cylinders for aerodynamic evaluation in wind tunnel experimentations. The study showed that the number of the cylinders representing the body parts can be simplified according to the body positions in different sports. The simplified human body represented with multiple cylindrical segments is illustrated in Figure 1(a). The body parts covered with fabrics can influence the aerodynamic behaviour by altering the air flow characteristics without affecting the body position. The air flow characteristics can also be influenced by varying angles of attack. Figure 1(b) shows the segmentation of a sprinter in running position. The breakdown of individual body parts clearly show that these parts can be treated as multiple cylinders with varied dimensions and positions (Chowdhury et al., 2009, 2010).

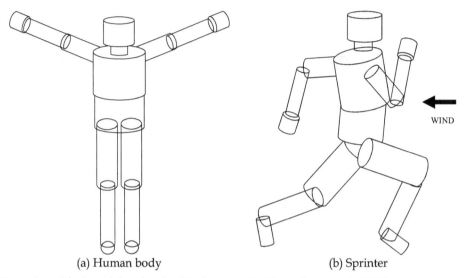

(a) Human body   (b) Sprinter

Fig. 1. Simplified model with cylindrical segments (Chowdhury et al., 2010)

Thus, the aerodynamic characteristics of these cylindrical body parts can easily be evaluated in wind tunnel testing under a range of angles of attacks and yaw angles representing real life body position in sporting actions using the articulating cylindrical methodology developed by Chowdhury et al. (2009, 2010). The study demonstrated that the cylindrical arrangements can be used to evaluate the aerodynamic properties of fabric features such as seam position, wale orientation and surface roughness with standard cylindrical geometry to quantify the aerodynamic properties more accurately.

In this study, the drag and lift characteristics of a number of sports fabrics were evaluated in various configurations together with the implications of this behaviour for the design of garments in such higher speed sports. Two sports were selected based on their speed ranges and characteristics. One is ski jumping where the speed is about 90 to 100 km/h. Another one is cycling where the speed is relatively lower, ranges from 30 to 60 km/h. For ski jumping, both lift and drag optimisation are important as the ski jumper become airborne during the in-flight phase. On the other hand, in cycling, the minimisation of aerodynamic drag is paramount. Full scale evaluation for ski jumping and cycling were also conducted. Full scale mannequin or human athlete test results were correlated with the standardised cylinder test data. Therefore, the focus of the research is on the comprehensive evaluation of aerodynamic behaviour of sports garments used in ski jumping and cycling sports to determine aerodynamic efficiency. No such study has been reported in the literature.

This chapter will describe a systematic methodology to design aerodynamically optimised sports garments starting from the wind tunnel experimentations of fabric sleeves with cylinder models to the full scale evaluation of prototype suits. Details studies were carried out for cycling and ski jumping suits, although this methodology can be applied to other higher speed sports based on their characteristics.

2. Methods, equipment and materials

2.1 Segmentation of athlete body

In this study, the segmentation of different body parts for ski jumper and cyclist were considered. Figure 2 shows the cylinder representation of different body parts of a ski jumper. Simplified model of ski jumper was modelled with 5 cylindrical shapes based on the ski jumping at in-flight position obtained for the field research results of 19th Olympic Winter Games in 2002 (Schmölzer & Müller, 2004). Cylinder 1 and 2 represent the arms, cylinder 3 and 4 represent legs and cylinder 5 represents the trunk of the ski jumper. It is clearly evident that these cylinders have different angle of attack as shown in the side view. Angles of different body parts of a ski jumper are indicated in Figure 3. Table 2 indicates the angular positions of arm and leg segments of a ski jumper at in-flight position.

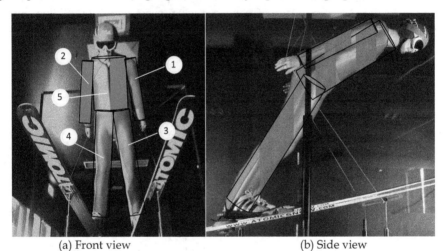

(a) Front view   (b) Side view

Fig. 2. Cylinder representation of different body parts of a ski jumper

Body Segment	Angle of Attack ($\alpha°$)	Rotation ($\psi°$)
Arm ($\alpha1$)	30	5.3
Leg ($\alpha3$)	45	30

Table 2. Angular position for leg and arm segments of a ski jumper

Similarly, cyclist body parts were also decomposed in to cylindrical geometry. Figure 4(a) shows the cylinder representation of different body parts of a cyclist. The body parts were considered only for those parts that are covered with fabric. Simplified model of a cyclist was modelled with 4 cylinders. Cylinder 1, 2, 3 and 4 represent the forearms, arms, trunk and thigh of the cyclist respectively. It is clearly seen that these cylinders have different

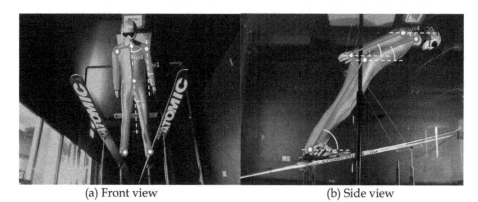

| (a) Front view | (b) Side view |

Fig. 3. Angles of different body parts of a ski jumper

angle of attack. Also the angles are varied in different cyclist positions. In competitive cycling, 2 main cycling positions are generally being adapted by the cyclist (Kyle et al., 2004; Lukes et al., 2005). Angle of attack of body parts varies depending on cycling position. Angles of different body parts of a cyclist are indicated in Figure 4(b). Table 3 indicates the angular positions of different body parts of a cyclist at 2 cycling positions.

| (a) Cylinder representation | (b) Angles of a cyclist body parts |

Fig. 4. Cylinder representation and angles of a cyclist body parts that are covered with fabric

Body segment angle	Road racing position	Time trial position
Thigh ($\alpha 1$)	0°- 115°	0°- 115°
Trunk ($\alpha 2$)	45°	0°
Arm ($\alpha 3$)	115°	105°
Forearm ($\alpha 4$)	115°	0°

Table 3. Angles of different body segment in two cycling positions

The rotational cylinder arrangement was used to measure the aerodynamic forces for cycling fabrics for different segments of the cyclist. Two samples were tested with a range of speed from 20 to 70 km/h with angles of attack at 30°, 45°, 60°, 105° and 115° which covers all the body parts angles at different cycling positions (see Table 3). Rotation angle (yaw) was set to zero for all configurations. Aerodynamic evaluation of both ski jumping and cycling fabrics was carried out with rotational cylinder methodology (Chowdhury et al., 2009) which will be discussed in the next subsection.

2.2 Fabric testing using cylinder testing methodology

In order to evaluate aerodynamic properties (drag and lift forces and their corresponding moments) of these cylindrical body parts under a range of positions, a wind tunnel experimental methodology was developed by Chowdhury et al. (2008, 2009, 2010) with standard cylinder with variable angle configurations to measure the drag and lift simultaneously at different yaw and angle of attack. Figure 5 shows the schematic of the experimental arrangement. It consists of a solid cylinder with 110 mm diameter and 300 mm length. The cylinder is connected with 6-componemt load cell through a steel strut. Fabric sleeves to be tested are wrapped against the cylinder. The arrangement was designed with a rotating mechanism to allow the cylinder to fix any angle from 30° to 150° relative to the wind direction and also the yawing can be done by using rotating table that allows the cylinder to be rotated and fixed at any angle from 0° to 360°. Figure 6 shows the angle of attack and yaw angle measurement in the experimental arrangement.

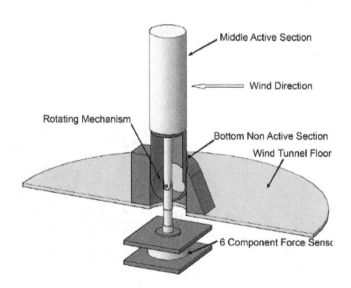

Fig. 5. Schematic of the experimental arrangement

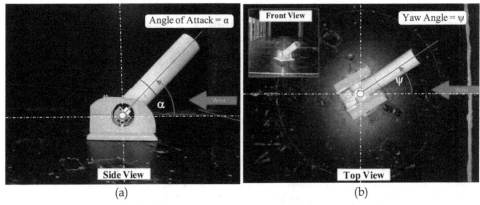

Fig. 6. Inclined cylinder arrangement in the wind tunnel

2.3 Full scale testing methodology

In order to measure the aerodynamic properties of ski jumping and cycling suits, two separate experimental arrangements and methodologies were used. Details of the experimental methodologies can be found in Chowdhury et al. (2010, 2011a, 2011b). For the ski jumping suits, an articulated mannequin (178 cm height) including all equipment (e.g., skis, boots, goggles and helmet) was used in wind tunnel testing for the replication of the ski jumper at in-flight position based on the field research results of 19th Olympic Winter Games in 2002 (Schmölzer & Müller, 2004). Figure 7 shows the experimental arrangement in the wind tunnel for ski jumping suit testing. Suits were tested at a fixed position with $\alpha=10°$, $\beta=20°$, $\gamma=160°$ and angle between two skis, V=30°.

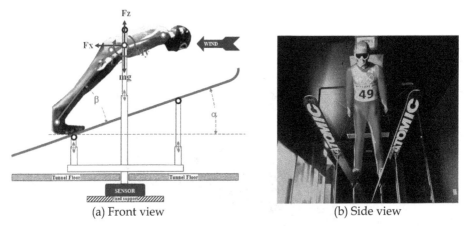

Fig. 7. Experimental arrangement in the wind tunnel for ski jumping suit testing

On the other hand, cycling suits were tested in the wind tunnel at road racing (upright) and time trial positions by using a professional cyclist. During the testing, professional racing bicycles along with appropriate helmets, and other accessories were used to replicate the real cycling as possible. For the road racing position, a professional road racing bicycle (Orbea) and a Giro Atmos road racing helmet were used as shown in Figure 8(b). Figure 8(b) shows the time trial position in which a professional time trial bicycle (Louis Garneau) and a Giro Advantage time trial helmet were used. It may be mentioned that all necessary documentations on ethics of human experimentation was undertaken from the RMIT University Ethic Committee prior conducting the experiments.

(a) Road racing position   (b) Time trial position

Fig. 8. Cycling suits testing in the wind tunnel

In order to reposition the experimental arrangement and other equipment for repeatable data acquisition, video positioning system was developed. Details of the video positioning system can be found in Chowdhury et al. (2011b).

2.4 Experimental facilities and equipment

The RMIT Industrial Wind Tunnel was used for this study. The wind tunnel is a closed return circuit wind tunnel that has a rectangular test section of 6 square meters with low turbulence (about 1.8%). The dimension of the test section is 2 m × 3 m × 9 m with a turntable to yaw suitably sized models. The tunnel used a 7-blade fan with the approximate diameter of 3 m, which is driven by a DC electric motor controlled by a tachometer mounted on the output shaft of the motor. The maximum wind speed in the test section is approximately145 km/h. A remotely mounted fan drive motor and acoustically treated turning vanes minimise the background noise and temperature rise inside the test section. Also the blockage ratio for the full scale test was less than 10%. A plan view of the tunnel is shown in Figure 9. The tunnel was calibrated before conducting the experiment.

The wind speeds inside the wind tunnel were measured with a modified National Physical Laboratory (NPL) ellipsoidal head pitot-static tube (located at the entry of the test section) which was connected through flexible tubing with the Baratron pressure sensor made by MKS Instruments, USA. Temperature, dynamic pressure, and upstream velocity inside the wind tunnel in real time are obtained from the wind tunnel control panel.

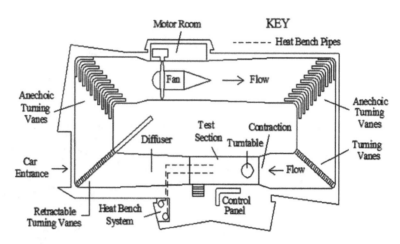

Fig. 9. A schematic of the wind tunnel

To measure the forces and moments in real time, a digital data accusation system was used. The system consists of a load sensor, connecting cable, PCI data card (12 bit) and data acquisition computer (Microsoft Windows 2000 compatible) with a data acquisition software. The experimental arrangements were connected through a mounting sting with the JR3 multi-axis load cell. The sensor was used to measure all three forces (drag, lift and side forces) and three moments (yaw, pitch and roll moments) at a time. Accuracy of the sensor is nominally 1% of full scale. Each data point is recorded for 30 seconds time average with a frequency of 20 Hz ensuring electrical interference was minimised. Multiple snaps were collected at each speed tested and the results were averaged for minimising the further possible errors in the raw experimental data which are further processed using Microsoft Excel spread sheet analysis software. Two M series JR3 sensors with load rating 200N and 1000N were used for the measurement of aerodynamic properties for cylinder and full scale testings respectively.

2.5 Materials

Three different fabrics were studied. Fabrics are generally made of yarn or thread that is a continuous twisted strand of wool, cotton or synthetic fibre with various stitch configurations. Figure 10 illustrates the structure of a common knitted fabric. Stitch density of fabric is expressed as the amount of courses and wales per centimetre. Course can be defined as the meandering path of the yarn through the fabric. Figure 10 shows the yellow path that defines one course. Wale is a sequence of stitches in which each stitch is suspended from the next as shown in Figure 10.

One ski jumping fabric material was studied with three wale orientations. Figure 11 shows the fabric orientation of the samples facing the wind. Wale direction is shown in Figure 11 with straight parallel lines. The figure shows that the wale directions of Sample#01, Sample#02 and Sample#03 are positioned at 0°, 45°, and 90° respectively with respect to wind direction. The position of seam of all 3 samples was placed at the back at 180° (opposite the wind flow direction) in order to avoid the flow disturbances due to the seam.

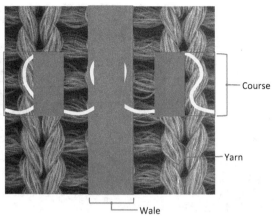

Fig. 10. Structure of a common knitted fabric stitch

Ski jumping fabric material consists of 5 layers. Outer and inner layer (thickness 0.5mm each) are knitted fabrics and the middle layers are (thickness 2 mm each) foam layers with an elastic perforated membrane at the centre layer. Other 2 knitted fabrics with average thickness of 0.5 mm were studied for cycling.

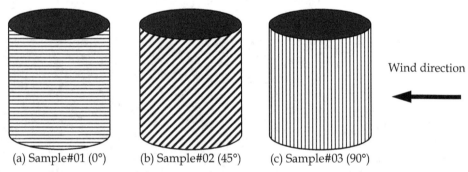

(a) Sample#01 (0°)   (b) Sample#02 (45°)   (c) Sample#03 (90°)

Fig. 11. Wale orientations of ski jumping fabric

For this study, a Field Emission Scanning Electron Microscope (model: FEI Quanta 200) was used for capturing fabric surface images with different magnification and Alicona Mex® software was used to generate 3D model of the scanned images for surface profile analysis. Figure 12 shows the 3D model of a fabric surface. Table 4 listed the surface measurement of the fabrics used in this study. Fabric sleeves with a single seam were fabricated for the cylinder such that each fabric had similar tensions when installed on the cylinder.

Fabric	Average height of fabric surface (µm)
Sample#01, 02 and 03	48.179
Sample#04	52.052
Sample#05	59.670

Table 4. Average height of fabric surface

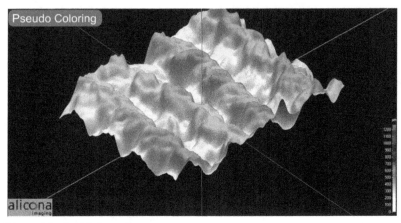

Fig. 12. 3D model of a fabric surface

Two ski jumping suits were fabricated for full scale testing in the wind tunnel. The suits are shown in Figure 13. Ski jumping suit is generally a single piece suit made of 13 panels with a zipper at the front. The panels are stitched together using seams. Both suits were made according to the body measurement of the mannequin used for full scale wind tunnel testing where seams were positioned at 90° with respect to the wind direction. Table 5 represents the configuration of the ski jumping suits.

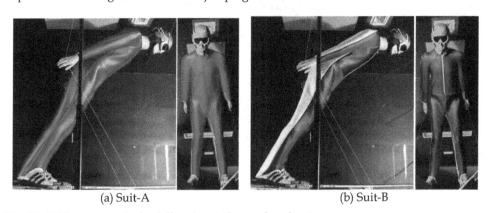

(a) Suit-A   (b) Suit-B

Fig. 13. Ski jumping suits for full scale wind tunnel evaluation

Parameters	Suit-A	Suit-B
Colour	Blue	Bottle green and white
Leg sleeve fabric	Sample#01	Sample#02
Arm sleeve fabric	Sample#01	Sample#03

Table 5. Configuration of the ski jumping suits

Two prototype cycling skin body suits (single piece) were also fabricated with different fabrics as shown in Figure 14. The black colour suit (Suit-1) is made of Sample#04 and the

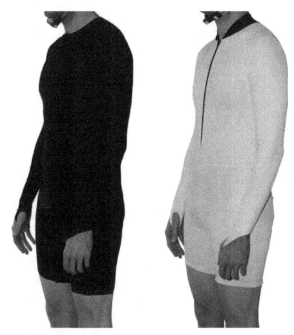

Fig. 14. Cycling suits for full scale wind tunnel testing

white colour suit (Suit-2) is made of Sample#05. Both suits were manufactured according to the body measurements of the cyclist and were tested in the wind tunnel in two cycling positions (road racing and time trial) using a professional cyclist including appropriate bicycles, helmets and other accessories.

3. Results and discussion

3.1 Fabric testing using cylinder methodology

The rotational cylinder methodology was used to evaluate the aerodynamic properties of the ski jumping fabric sleeves (Sample#01, Sample#02 and Sample#03) for leg ($\alpha = 45°$ and $\psi = 30°$) and arm ($\alpha = 30°$ and $\psi = 5.3°$) segments. Tests were conducted at a range of wind speeds (80 km/h to 110 km/h with an increment of 10 km/h) and the drag (D) and lift (L) forces are measured and converted to their dimensionless quantity drag coefficient (C_D) and lift coefficient (C_L). The C_D and C_L were calculated by using the following formulae:

$$C_D = \frac{D}{\frac{1}{2}\rho V^2 A} \tag{1}$$

$$C_L = \frac{L}{\frac{1}{2}\rho V^2 A} \tag{2}$$

where, ρ, V and A are the density of the air, wind speed and frontal projected area respectively.

Figures 15 and 16 show the results for the evaluation of ski jumping fabric for arm and leg segments respectively. The results show the C_D, C_L and L/D variations with speed (km/h).

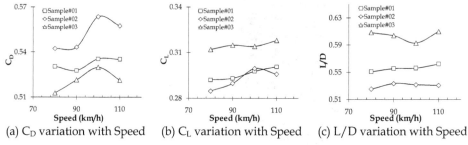

(a) C_D variation with Speed   (b) C_L variation with Speed   (c) L/D variation with Speed

Fig. 15. Evaluation of ski jumping fabric at arm position (α = 30° and ψ = 5.3°)

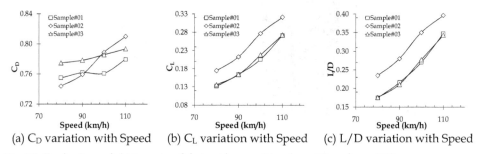

(a) C_D variation with Speed   (b) C_L variation with Speed   (c) L/D variation with Speed

Fig. 16. Evaluation of ski jumping fabric at leg position (α = 45° and ψ = 30°)

In flight aerodynamics, the quality of lifting components is evaluated by the ratio of L/D (Meile et al., 2006). From the cylinder test results, it is clearly evident that Sample#03 has maximum L/D value than any other configurations within the speeds range from 90 to 100 km/h. Similarly, Sample#02 exhibits maximum L/D value for the leg configuration. Thus, it is believed that similar aerodynamic advantage can be obtained in full scale ski jumping suit by implementing the arm and leg segments of the suit according to these configurations.

In order to understand the effect of cylinder test results, two full scale suits were manufactured for full scale aerodynamic evaluation. One of these two suits (Suit-B) was made according to with maximum L/D ratio obtained from the cylinder testing methodology and other one (Suit-A) was made with Sample#01 for both the leg and arm segments. Tables 6 and 7 summarise the aerodynamic parameters for arm and leg segments of the ski jumping suit at a speed range from 90 to 100 km/h.

Parameters	Sample#01	Sample#03	Changes (%)
CD	0.532	0.526	-1.1%
CL	0.296	0.314	+6.4%
L/D	0.556	0.598	+7.6%

Table 6. Fabric comparison for arm segment

Parameters	Sample#01	Sample#02	Changes (%)
CD	0.761	0.774	+1.7%
CL	0.185	0.245	+32.2%
L/D	0.243	0.315	+29.7%

Table 7. Fabric comparison for leg segment

For the comparison of cylinder test results with the full scale evaluation, Sample#01 was taken as the base configuration. The changes (%) in aerodynamic properties were estimated in comparison with the base configuration (Sample#01) for both arm and leg segments. Table 6 shows that the C_D is decreased by 1.1% but C_L is increased by 6.4% for arm position. Also, the glide ratio (L/D) is increased about 7.6% which is an aerodynamic advantage. Similar aerodynamic benefit is also observed at leg position with an increase of L/D ratio about 30% (see Table 7). Full scale testing of these suits was carried out and the results will be discussed in next subsection.

In order to compare the cylinder test results with full scale evaluation, two cycling fabric samples — Sample#04 (relatively smooth surface) and Sample#05 (relatively rough surface) were selected for this study. The fabric samples were tested with a range of angles of attack starting from 30° to 115°. These angles were selected based on the cyclist's body position during different cycling positions as elaborately discussed in Section 2. The impact of surface roughness in relation to different body position were compared with cylinder test results with full scale testing which will be discussed in next subsection. Figure 17 shows the C_D variation with speed for 2 cycling fabrics at different angles of attack.

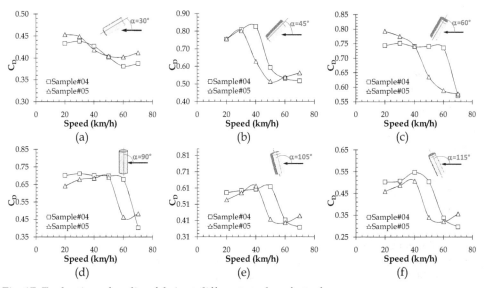

Fig. 17. Evaluation of cycling fabric at different angles of attack

Results indicate that relatively rough fabric (Sample#05) undergoes flow transition earlier than relatively smooth fabric (Sample#04) at high angle of attack (α = 45° to 115°). On the other hand, Sample#04 has lower C_D value (approximately 5% less) at low angle of attack (α = 30°) and higher speed (when speed is 40 km/h or over). It is clearly evident that with an increase of angle of attack, the rough fabric reduces drag more efficiently. However, when the body become more streamlined (at α<45°) then the smooth fabrics can be used to reduce drag more efficiently. In the time trial position, the cyclist body become streamline than road racing positions. Therefore, the smooth fabric has the potential to provide aerodynamic advantage at this position. But when the cyclist body is less inclined as in road racing (upright) position when the trunk, arm and thigh angles are more than 30°, then rough fabric has the potential to reduce overall drag at the low speed ranges from 30 to 40 km/h.

3.2 Full scale evaluation of sports garments

Two ski-jumping suits were tested for the full scale aerodynamic evaluation at two different positions. Frontal areas of the ski jumping experimental arrangement including the ski jumper at both positions were measured using the projected frontal area measurement system described in Chowdhury et al., 2011a. Drag and lift forces were measured at a range of wind speeds from 80 to 110 km/h. Drag and lift forces were converted to their non dimensional drag and lift coefficient. Figure 18 shows the full scale testing of ski jumping suits for the C_D, C_L and L/D variations with speed with error bars indicating the standard deviation of the experimental data.

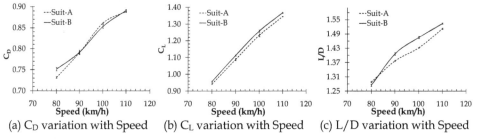

(a) C_D variation with Speed   (b) C_L variation with Speed   (c) L/D variation with Speed

Fig. 18. Full scale testing of ski jumping suits

Results show that there is little variation in C_D values between two suits in both positions within the speed range from 90 to 100 km/h. Suit-B exhibits more C_L value in both positions than Suit-A. Also L/D values of Suit-B are higher than that of Suit-A. Thus a little aerodynamic benefit is noted with Suit-B. With Suit-B, the Drag is decreased by 0.4% but the lift is increased by 2.2% with respect to Suit-A. The resultant jump length can be increased by about 4.44 m by using the simulation approach formulated by Muller & Schmolzer (2002).

Two cycling suits were manufactured for the full scale testing. Suits were tested at two widely used cycling positions: road racing and time trial. Projected frontal areas of the cyclist at different positions were measured using the projected frontal area measurement system described by Chowdhury et al., 2011b. Drag forces were measured under a range of wind speeds starting from 20 to 70 km/h with an increment of 10 km/h. Drag forces were converted to non dimensional drag coefficient (C_D). Figure 19 shows the C_D variation with speed at road racing and time trial position.

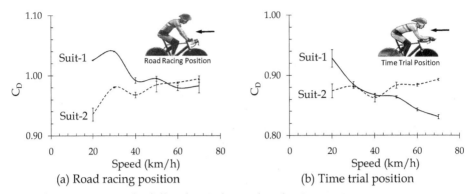

Fig. 19. Ski jumping suits for full scale wind tunnel evaluation

The road racing (upright) position is used by the cyclist generally during long distance road racing. In this position, the frontal area is comparatively higher than time trial positions. As a result, the cyclist experiences more aerodynamic drag. The speed range for upright position is from 30 to 40 km/h. Suit-2 is made of relatively rough fabric (Sample#05) and has lowest C_D values from 30 to 50 km/h as shown in Figure 19(a). Therefore, Suit-2 provides more aerodynamic benefits at this cycling position and speed range compared to Suit-1 which is made of relatively smooth fabric (Sample#04). Results show that around 2.1% drag reduction is possible with Suit-2 at upright position between 30 to 40 km/h.

On the other hand, in time trial position is used by the cyclist generally during the time trial cycling event. At this position, the cyclist tries to achieve the minimum projection frontal area in order to minimise the aerodynamic drag. This is the most streamlined position used by the cyclist. Figure 19(b) shows the C_D variation with speed at time trial position. The speed range for the time trial position is generally 50 to 60 km/h. Suit-1 has lowest C_D value within this speed range.

3.3 Comparison of results

In ski jumping, it was predicted earlier with the cylinder results that Suit-B would be more aerodynamically efficient than Suit-A. Cylinder results indicate an increase of L/D ratio approximately 30% for the leg segment and 8% for the arm segment compared to the base configuration (Sample #01). Full scale results shows that Suit-B has approximately 3% increase of L/D compared to Suit-A which is made according to base configuration. Therefore, it is evident that cylinder results has similar trend with full scale results. However, the magnitude of the aerodynamic benefit is different. Cylinder test results are more specific to the fabric properties. Full scale testing is more complicated compared to simplified cylindrical methodology.

In cycling, the average drag reduction was calculated by taking Suit-1 as the base for comparison. The average drag reduction with Suit-2 was found at road racing (upright) position by 2.1% approximately. But the drag increased by 1.5% at the time trial position at low speed (below 50 km/h). The average reduction of drag for the time trial positions was around 21% compared to the upright position. Suit-2 has the highest reduction of drag in road racing (upright) position where generally the cyclist's body is not very streamlined. Suit-1 performs better than Suit-2 at time trial position at high speeds (over 50 km/h).

Cylinder test results indicate that the smooth fabric possesses 5% less drag than rough fabric at low angle of attack ($\alpha < 45°$) at speed ranges from 50 to 60 km/h. Full scale results indicate 2.1% drag reduction with Suit-1 at time trial position compared to Suit-2. Cylinder results also showed that the rough fabric undergoes the flow transition earlier than the smooth fabric when angle of attack is more than 45°. Therefore, at low speeds (from 40 to 50 km/h), C_D values of the rough fabric are lower than that of the smooth fabric. The average drag reduction was calculated around 3.3% with Suit-2 compared to Suit-1. Therefore, it is evident that full scale test results compared well with cylinder results.

However, the magnitudes of the aerodynamic forces are less in cylinder testing. Cylinder results are more specific to fabric features (e.g., surface profile) of the suit compare to the full scale testing. The full scale testing is less sensitive as the high magnitude of force is measured and also the aerodynamic interference is more. Therefore, the cylinder test is more appropriate as the flow transition due to the surface roughness can be observed with cylinder testing. On the other hand, no flow transition was observed in the full scale testing for ski jumping and cycling which agreed well with published data (Brownlie, 1992). Brownlie tested several fabrics with cylinder model and garments with full scale models for cycling and sprint. The study showed that flow transition was only observed with cylinder model but no flow transition was noted with full scale model. Brownlie (1992) suggested that peripheral parts (arm, leg, etc.) of human athlete prevented the flow separation. In this study, complex flow structures were observed around the ski jumping and cycling using the flow visualisation technique which is not included in this study. However, on field test can be another way for further verification of cylinder test results.

4. Conclusion

Cylinder results have good agreement with the full scale testing both for ski jumping and cycling garments. In this study, it was estimated about 4.44 m increase of the jump length which could be a decisive advantage for a ski jumper. The results also demonstrated that the cycling suit should be selected depending on the cycling position and speed range in order to take the aerodynamic advantage. As the position in the world class competitions are decided with a fraction of time difference, apart from the athletic performance, an efficient sport garment can enhance the overall performance of the athlete. Depending on the nature of the sport, this methodology can be used as a basic design tool to optimise or select proper parameters for the betterment of the outcome. Prior to making a full-scale suit, this methodology can be an essential tool to investigate the performance of individual parts of the body. Thus, it will be easier to manufacture the aerodynamic prototype suit. The practical implementation of knowledge from this study can be applied not only in ski jumping and cycling but also other speed sports.

5. Acknowledgment

The author is highly grateful to Dr Firoz Alam, Prof. Aleksandar Subic, Prof. David Mainwaring and Dr Margaret Tate of RMIT University for their guidance and assistance. The author thanks Mrs Dorothy Forster for fabricating the fabric sleeves, ski jumping and cycling suits and also thanks Mr Jordi Beneyto-Ferre for his active participation in the wind tunnel testing. Author acknowledges the support from the School of Applied Sciences, RMIT University for providing the articulated mannequin.

6. References

Brownlie, L. W. (1992). Aerodynamic Characteristics of Sports Apparel, PhD Thesis, University of British Columbia, Canada

Capelli, C.; Rosa, G.; Butti, F.; Ferreti, G., Veicsteinas, A. & Di Prampero, P. E. (1993). Energy cost and efficiency of riding aerodynamic bicycles, *European Journal of Applied Physiology*, Vol.67, pp. 144-149

Chowdhury, H.; Alam, F. & Mainwaring, D. (2011a). Aerodynamic study of ski jumping suits. *Procedia Engineering*, Vol. 13, pp. 376-381

Chowdhury, H.; Alam, F. & Mainwaring, D. (2011b). A full scale bicycle aerodynamics testing methodology. *Procedia Engineering*, Vol. 13, pp. 94-99

Chowdhury, H.; Alam, F. & Subic, A. (2010). Aerodynamic Performance Evaluation of Sports Textile. *Procedia Engineering*, Vol.2, N0.2, pp. 2517-2522, Elsevier, UK

Chowdhury, H.; Alam, F. & Subic, A. (2010). An Experimental Methodology for a Full Scale Bicycle Aerodynamics Study, In: *Proceedings of the First International Conference on Mechanical, Industrial and Energy Engineering (ICMIEE2010)*, Paper MIE10-124, ISBN: 978-984-33-2300-2, Khulna, Bangladesh

Chowdhury, H.; Alam, F.; Mainwaring, D.; Beneyto-Ferre, J.; Forster, D.; Tate, M. & Subic, A. (2010). Experimental Evaluation of Ski Suit Performance, In: *Proceedings of the 17th Australian Fluid Mechanics Conference (AFMC2010)*, Paper 173, ISBN: 978-0-86869-129-9, Auckland, New Zealand

Chowdhury, H.; Alam, F.; Mainwaring, D.; Subic, A.; Tate, M.; Forster D. & Beneyto-Ferre, J. (2009). Design and Methodology for Evaluating Aerodynamic Characteristics of Sports Textiles, *Sports Technology*, Vol.2, No.3-4, pp. 81-86, John Wiley and Sons Asia Pte Ltd.

Chowdhury, H.; Beneyto-Ferre, J.; Tate, M.; Alam, F.; Mainwaring, D.; Forster, D. & Subic, A. (2009). Effects of Textile and Garment Design on Aerodynamic Characteristics Applied to Cycling Apparel, In: *The Impact of Technology on Sport III*, Paper no.: P-131, ISBN: 978-1-921426-39-1, Hawaii, USA

Chowdhury, H.; Alam, F.; Mainwaring, D.; Subic, A.; Tate, M. & Forster, D. (2008). Aerodynamic Testing Methodology for Sports Garments, In: *Proceedings of the 4th BSME-ASME International Conference on Thermal Engineering*, Paper no.: AU-04, ISBN: 984-32-3815-X, Dhaka, Bangladesh

Di Prampero, P. E.; Cortili, G.; Mognoni, P. & Saibene, F. (1979). Equation of motion of a cyclist, *Journal of Applied Physiology*, Vol.47, pp. 201-206

Dias, T. & Delkumburewatte, G. B. (2008). Changing porosity of knitted structures by changing tightness, *Fibers and Polymers*, Vol.9, No.1, pp. 76-79

Grappe, F; Candau, R; Belli, A & Rouillon, J. D. (1997). Aerodynamic drag in field cycling with special references to the Obree's position, *Ergonomics*, Vol.40, No.12, pp. 1299-1311

Hoerner, S. F. (1952). Aerodynamic Properties of Screens and Fabrics, *Textile Research Journal*, Vol.1, No.22, pp. 274-279

Kyle, C. R.; Brownlie, L. W.; Harber, E.; MacDonald, R. & Norstrom, M. (2004). The Nike
    Swift Spin Cycling Project: Reducing the Aerodynamic Drag of Bicycle Racing
    Clothing by Using Zoned Fabrics, In: *The Engineering of Sport 5, Vol. 1*, International
    Sports Engineering Association, UK
Lukes, R. A.; Chin, S. B. & Haake, S. J. (2005). The understanding and development of
    cycling aerodynamics, *Sports Engineering*, Vol.8, No.2, pp. 59-74
Meile, W.; Reisenberger, E.; Mayer, M.; Schmölzer, B; Müller, W. & Brenn, G. (2006).
    Aerodynamics of ski jumping: experiments and CFD simulations, *Experiments in
    Fluids*, Vol. 41, No.6, pp. 949–964
Moria, H.; Chowdhury, H.; Alam, F.; Subic A.; Smits, A. J.; Jassim, R. & Bajaba, N. S. (2010).
    Contribution of swimsuits to swimmer's performance, *Procedia Engineering*, Vol.2,
    No.2, pp. 2505-2510, Elsevier, UK
Muller, W. (2008). Computer simulation of ski jumping based on wind tunnel data, *Sports
    Aerodynamics*, pp. 161-182
Müller, W. & Schmölzer, B. (2005). Individual Flight Styles in Ski Jumping: Results
    Obtained During Olympic Games Competitions, *Journal of Biomechanics*, Vol.38, No.
    5, pp. 1055-1065
Oggiano, L.; Troynikov, O.; Konopov, I.; Subic, A. & Alam, F. (2009). Aerodynamic
    behaviour of single sport jersey fabrics with different roughness and cover factors,
    Sports Engineering, Vol.12, No.1, pp. 1-12
Schmölzer, B. & Müller, W. (2004). Individual flight styles in ski jumping: results obtained
    during Olympic Games competitions, *Journal of biomechanics*, Vol.38, pp. 1055-1065
Spencer, D. J. (2001). *Knitting Technology: a comprehensive handbook and practical guide*,
    Woodhead Publishing, Cambridge, England

2

Human-Powered Vehicles – Aerodynamics of Cycling

José Ignacio Íñiguez, Ana Íñiguez-de-la-Torre
and Ignacio Íñiguez-de-la-Torre
Universidad de Salamanca, Departamento de Física Aplicada
Spain

1. Introduction

Human-powered transport can be defined as that type of transport that only uses the human muscle power. When considering non-vehicular human-powered transport, it should to be indicated that it exists from the beginning of the human history (walking, running, swimming or climbing to trees). On the contrary, if we refer to vehicular transport we must necessary mention the wheel invention. It is superfluous to remark that the wheel is the most essential element of any form of vehicular land transportation and perhaps the greatest mechanical invention of the human civilization. Although archaeologists say that the wheel was discovered around 8000 BC in Asia, today is generally accepted that the oldest wheel known was discovered around the year 3500 BC in Sumeria, part of modern day Iraq. It was manufactured of slats of wood linked together. The wheel allowed the people to travel with greater speed and efficiency than walking: In a first step with the help of animal-powered rolled chariots and from the 19th century in human-powered vehicles also.

The most efficient human-powered vehicle is obviously the bicycle. Modern technology allows manufacturing high efficiency bicycles permitting to extraordinarily increase the human muscle power effectiveness. During the last century the development of petrol, gasoil and electric engines have increased load capacity and speed in many kinds of vehicles in such a manner that human-powered transport (mostly the bicycle) has been clearly relegate. Fortunately, in recent days we assist to a renaissance of cycling due to a variety of reasons as can be low cost, physical exercise, healthful sport, leisure or ecological and sustainable transport in the center of our populous cities.

It is interesting to remark that today is frequent to find papers devoted to the study of the stability of the bicycle. It seems clear that the gyroscopic contribution is not the main assistance to guarantee the stability of a bicycle, but however the problem is not completely solved yet [Jones, 1970; Cleary et al., 2011; Kooijman et al., 2011].

The scarcity of available power in human-powered machines is the main handicap for its effective development. In order to be conscious of this fact we remember that any motorized conventional vehicle has power ranging from some kW to a few hundreds of kW, while the useful power of a human is about some hundreds of W. Let's say that for a person of 70 kg

Fig. 1. The "High-Wheel" bicycle, 1870

the metabolic consumption at rest corresponds to 80 W and when sitting (reading, writing or watching TV) he requires a surplus of 40 W, while to make usual home activities power ranging from 50 to 100 W must be added. Walking at 1 m/s speed develops 150 W of additional power, i.e. a total of 230 W approximately is required. When running at 4-5 m/s speed or speedily climbing a mountain in bicycle the power requirement is near 0.6 to 1 kW, which constitutes a quantity that only very well trained sportsmen can maintain during a few minutes. Higher powers can be exclusively developed during a few seconds, as a sprint (running or cycling) with peak power about 2 kW, or a fraction of a second, as can be an athletic jump with 4 kW [Lea & Febiger , 1991; Bent, 1978; Eriksen et al. 2009].

A simple comparison of these figures with the energy consumption of our civilization results very instructive. The mean value of the amount of petroleum in 2008 was 4.5 barrels per year and person, but varies a lot from country to country. That year, in the USA, people used about 25 barrels of oil per capita whereas in Nigeria or Phillipines the amount was only 2 or 3 barrels. Taking into account that a barrel of oil is equivalent to 6.1 GJ, it results very easy to obtain that a USA citizen spent in 2008 a mean power of 4850 W coming from fossil fuel (irrespectively of others energy sources as carbon, gas, nuclear, hydroelectric, solar, eolic...), which roughly corresponds to a quantity 50 times greater than the mean power (100 W) of a person [Indexmundi, 2011].

2. Different types of human-powered vehicles

There are a lot of human-powered vehicles. Some of them are designed for land transportation (bicycle, monocycle, tricycle, quadracycle, recumbent bicycle, tandem bicycle, taxi-cycle, skateboard, ice skating, skiing, handcar...), others for travel in water (canoe, gondola, pedal boat, hydro-bike, galley, submarine...) and others to fly (aircrafts, helicopters...). With the exception of research or recreational objectives, human-powered water and air vehicles must be considered of very minor importance when referring to useful transportation.

Anyway is interesting to emphasize that on 1988 a team of MIT students succeeded in a project called Daedalus (exclusively human-powered aircraft). It flew across the Mediterranean Sea from the Greek island of Crete to just a few meters from the coastline of the island of Santorini, using a set of bicycle pedals and a transmission chain (see Figure 3). Made largely of carbon-fiber composite and Mylar, it weighed just 31 kg. On its record flight, Daedalus travelled 115 km during 3 h 54 min across the sea before being buffeted by winds, breaking its tail spar and crashing into the waves just 7 meters offshore from its destination. The pilot, champion bicyclist Kanellos Kanellopoulos, could swim to shore unhurt. An identical craft used in the initial tests is on display at Boston's Museum of Science. The flight set the all-time records for duration and distance of a human-powered flight, handily beating the previous record of just under 36 km set by Gossamer Albatross in a crossing

Fig. 2. Modern propulsion technology in water transportation

Fig. 3. Daedalus human-powered aircraft (Massachusetts Institute of Technology, Photo/NASA/Beasley)

of the English Channel in 1979. Some technical characteristics of Daedalus are: Wingspan 34 m, weight 32 kg, pilot weight 68 kg, fuel (water and sugar) 4 kg, speed 25 km/h, useful power of the pilot at the takeoff 900 W, at stationary regime 240 W, corresponding to a total power at stationary regime near 900 W (efficiency 27 %), amount of heat dissipation 660 W (600 via perspiration and 60 by radiation and convection). Finally, the fuel consumption was merely 1 liter of water with sugar 10 % and salts 0.1 % per hour [Chandler, 2008; Wikipedia, 2011].

Returning to less sophisticated transportation vehicles, it is clear that, as it was stated above, the most efficient human-powered land vehicle is the bicycle. It is also more efficient (up to 5 times) than walking. With energy amounts of 100 kJ a bicycle and rider can cover more than 5 km on a horizontal road at 20 km/h speed while a car can only travel about 200 m. Today the number of bicycles in the world is higher than one billion, which constitutes a good prove of its efficiency. Besides of the conventional upright bicycle, the recumbent bicycle is more efficient on horizontal and downing hills roads, as corresponds to its better aerodynamics. Electric bicycles, designed for one person and small load capacity, provide convenient local and short distance transportation. They work on the basis of power assistance; the electric engine helps the rider when pedaling. Power and speed in electric bicycles are limited to about 250 W and 25 km/h respectively. Electric bicycles are clean, quiet and offer advantages at very low cost instead acquiring an additional little car.

Fig. 4. Modern time-trial bicycle

Fig. 5. A recumbent bicycle

Fig. 6. A foldable electric bicycle

3. Efficiency of muscle power in human powered land vehicles: The bicycle

The high efficiency of a bicycle allows travelling long distances with very low energy cost. A healthy man can develop a useful power of 100 to 150 W and continuously travel near 25 km/h speed on a horizontal path, depending on the equipment, position of the rider and type of road. Instead, well-trained amateur cyclists can produce 300 W during an hour while first-class cyclists develop power as high as 1500 W during a few seconds or more than 300 W for several hours.

The power and speed that a cyclist can develop increase with the muscle mass, which is dependent on the weight and the body fat percentage. When considering a horizontal road, as we will show in the next paragraph, the main contribution to required power in cycling is related to overcome the aerodynamic drag. Taller cyclists will present a major wind resistance needing to produce more power while more compact riders will spend less power. This is the reason why it is no frequent to find first-class cyclists with thin and tall physic complexion. Remember that while the frontal area is quadratic with the linear size of the cyclist, the power (the muscle mass) is cubic, as we will see below. Tall and heavy riders have advantage when downing hills or in horizontal time trials and perhaps in the sprint and, however, short and light cyclists are benefit when climbing hills. Therefore, the optimum weight for an all-terrain cyclist is about 70 kg whereas for a sprinter is around 90 kg [Padilla et al., 2000].

It is to be remarked that the work developed by a muscle is equal to the force that exerts multiplied by the displacement. The force is proportional to the cross-sectional area of the muscle (the number of muscle fibers) whereas the displacement is proportional to the length of the muscular contraction. Consequently, the work developed by the muscle is proportional to its volume, its muscle mass. This is therefore the reason why the index of power by mass unit (W/kg), so used in any text of physiology of the sports, acquires all their importance. Here we can see a set of data corresponding to perhaps the most legendary climb of the *Tour de France*, the Alpe d'Huez. An ascent of 1073 m with an average gradient of 7.9 % and 21 emblematic curves. A well-trained cyclist of 75 kg and a bicycle about 8 kg cycling at 300 W would ascend at 12 km/h and would take in crowning it one hour and five minutes. Until the end of the 1990 the Alpe d' Huez was climbed between 45 and 42 minutes, i.e. at speed ranging from 17 to 19 km/h. For instance, P. Delgado with 64 kg raised it in 1989 at 18.6 km/h, developing about 380 W. Between years 1990 and 1997 there was a revolution that began with M. Indurain that rose in 1995 at 19.9 km/h, showing the record of 470 W for his 80 kg during 39 minutes. M. Pantani, 57 kg, destroyed this record in 1997 with an average speed of 20.9 km/h during 37.5 minutes. More recently the Alpe d' Huez is being climbed at speed below 20 km/h as F. Schleck, of 67 kg, which developed a mean power of 407 W ascending at a speed of 19.3 km/h and took little more than 40 minutes in the year 2006. It is very simple, the less weight, the less power the cyclist need to go at the same speed; as higher is the ratio W/kg of the cyclist higher is also his possibility to win [Cyclingforums, 2011].

The way in which the human body produces energy is very similar to the behavior of a fuel cell. In fact, the own human body can be considered as the oldest and most refined fuel cell the man uses, where foods are catalytically oxidized in an electrolyte (blood) to produce the necessary power for the whole physiological requirements. The chemical energy residing in the chemical bonds of fuel molecules is directly converted into other forms of useful energy without employing any intermediate thermal cycle.

Although this form to produce energy is very different from the case of the conventional thermal machines (Otto or Diesel engines), the mean efficiency of both systems is very similar ranging from 20 to 30 % [Sonawat et al., 1984].

The rest of the power is lost in the form of heat and, in the case of the cyclist, in non-useful power developed during pedaling. Then, the refrigeration as well as the hydration are very important. With the aim to improve the way in which the cyclist does his pedaling, several technological advances have been implemented years ago. Doubtless, the most important is the use of a variable gear ratio. Modern derailleur transmissions are 2-3 % more efficient than hub gear transmissions. With the purpose of guarantee the possibility of maintain a constant pedaling cadence near 85 rpm, modern competition bicycles use two or three chain-wheels and eleven rear sprockets. The correct election of the crank length is also important as well as the minimization of the waste power during time that pedals are at the top and bottom of their circular trajectory and the torque they produce is almost null (the dead spots). Use of non-circular pedaling, elliptical chain-wheels or more complex transmissions that allow controlling a variable relation between the angular speeds of pedals and chain-wheels during every turn, the well-known Rotor-Bike, are today frequent [Wilson, 2004; Rotorbike, 2011].

4. The aerodynamics of cycling

Some important aspects concerning the aerodynamics of cycling will be here summarized. While the wind resistance or aerodynamic drag is a well-known concept to all the people and particularly to cyclists when travelling at moderate speed, some other aerodynamic phenomena are less well-known. There are two main issues to be here addressed. The first one corresponds to the non-neutral effect of the sidewind while the other one is concerning with the flow effects that appear when cyclists are riding one behind the other and the consequent saving of energy the riders experience, including the cyclist who travels in first position.

It is well established that the presence of wind is crucial for the practice of numerous outdoor sports. Among them, the case of cycling competition is especially important because the speeds that are developed are perfectly comparable with moderate or strong winds, giving rise to a great variability in race times. Conversely, the sidewind seems to behave as neutral in the race, and nevertheless all the cyclists know the difficulties that it causes. Here we will show that the sidewind also produces an appreciable braking as a consequence of the quadratic dependence of the aerodynamic drag force on the air speed.

On the other hand, cycle races, in which speeds of up to 15 ms^{-1} are frequent, offer great opportunities to appreciate the advantage of travelling in a group. We present a brief analysis of the aerodynamics of a cycling team in a time-trial challenge, showing how each rider is favored according to his position in the group. We conclude that the artificial tailwind created by the team also benefits the cyclist at the front by about 5 %. Also in this area, humans imitate nature. When seasonal journeys take place in nature, birds and fishes migrate in groups. This provides them not only with security but also a considerable saving of energy. The power they need to travel requires overcoming aerodynamic or hydrodynamic drag forces, which can be substantially reduced when the group travels in an optimal arrangement.

5. Cycling in the wind

The total force, F_T, that a cyclist exerts at a constant speed corresponds to three non-collinear quantities: the rolling resistance, F_R (necessary to overcome mechanical friction), aerodynamic drag force, F_D (to counteract the resistance force due to displacement through the air) and weight force, F_W (which allows displacement on a non-horizontal road).

The rolling force, opposing the direction of motion, is proportional to the friction coefficient, μ, and the normal component of the weight of the rider and bicycle:

$$F_R = \mu g (M + m) \cos \sigma \qquad (1)$$

where $g=9.8$ ms^{-2} represents gravitational acceleration; M and m correspond respectively to the mass of the rider and the bicycle, and σ represents the slope of the road.

This rolling force can be considered to be independent of the speed, and for a conventional racing bicycle with narrow tires (high pressure of about 0.7 MPa) rolling on a road in good conditions, the friction coefficient takes values close to 0.004 [di Prampero et al., 1979].

The aerodynamic drag force, in the direction of the air velocity relative to the cyclist, $V_{rel} = W - V$, (see figure 7), can be written as [Landau & Lifshitz, 1982]:

$$F_D = \frac{1}{2} \rho C_D A V_{rel}^2 \qquad (2)$$

where the air density, ρ, at the standard temperature and pressure takes values of around 1.2 kgm^{-3}. C_D is the drag coefficient, whose value is obtained as a function of the Reynolds number that we shall take as a constant for our purpose, irrespective of the angle between the directions of motion and wind. The projected area of the rider and bicycle, A, is also a function of the angle between the directions of motion and wind. The contribution of the bicycle is small and increases with the sidewind, while that corresponding to the cyclist is greater but diminishes with that wind. We can therefore disregard that dependence by

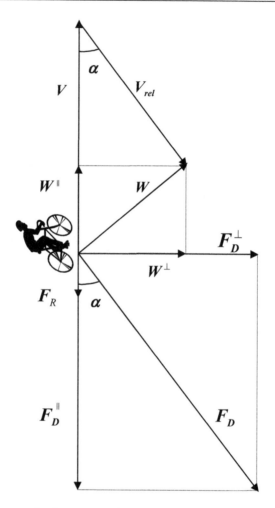

Fig. 7. Vectorial diagram of velocities and forces involved in cycling in the wind

taking A as a constant. According to several authors, and considering a mean sized cyclist, wearing cyclist clothing, rolling alone, and alternating his position on the bicycle between the top and the bottom of the handlebar, equation (2) can be written as [Wilson, 2004]:

$$F_D \approx 0.24 V_{rel}^2 \tag{3}$$

where the constant $\rho C_D A/2 \approx 0.24$ is expressed in SI units, Ns²m⁻². In this case, for our well-trained amateur cyclist, and assuming an air density of 1.2 kgm⁻³, the product $C_D A$ takes a value around 0.4 m², but it is to be noted that this quantity is dramatically sensitive to small variations in the cyclist size and position on the bicycle. In table 1, a set of results on experimental drag studies on the aerodynamics of cycling are presented [Gross et al., 1983, Wilson, 1997].

Bicycle type and rider position	Drag coefficient, C_D	Frontal area (m²)	$C_D A$ (m²)	Force to overcome the air drag at 10 ms⁻¹ (N)
Upright commuting bicycle	1.15	0.55	0.632	34.5
Road bicycle and amateur cyclist	1.0	0.4	0.4	22
Competition bicycle and well-crouched rider	0.88	0.36	0.32	17.6
Road bicycle with simple fairing	0.52	0.55	0.29	15.7
Moser bicycle	0.51	0.42	0.214	11.8
M5 full faired low racer	0.13	0.35	0.044	2.4

Table 1. Several data on aerodynamic of cycling

The effect of the position and geometry of the rider (prone or supine) and bicycle is extraordinary, but the use of high technology full fairings in recumbent bicycles is astonishing. It is possible reducing the required force to overcome an air resistance corresponding to a speed of 10 ms⁻¹ in more than one order of magnitude! Today, *The International Human Powered Vehicle Association* verifies the official speed records for recumbent bicycles. There are several well-defined modalities, the fastest of which is the "flying 200 m", a distance of 200 m from a flying start with a maximum allowable tailwind of 1.66 ms⁻¹. The current record is 133 kmh⁻¹, set by Sam Whittingham of Canada in a fully faired Varna Diablo front-wheel-drive recumbent low racer bicycle designed by George Georgiev [HPVA, 2011].

Finally, the weight force (opposing/parallel to the direction of motion) takes the form:

$$F_W = g(M + m)\sin\sigma \qquad (4)$$

where the slope of the road, σ, is positive/negative for uphill/downhill inclinations. With the aim of placing emphasis on the effect of wind, we shall consider a horizontal road ($\sigma = 0$). Assuming a mass of $(M + m) = 81$ kg, and according to figure 1, we can write the two components of the total force (parallel, $F_T^{||}$, and perpendicular, $F_T^{\perp}$, to the motion):

$$F_T^{||} = 3.2 + 0.24 V_{rel}^2 \cos\alpha = 3.2 + 0.24 V_{rel}^2 \frac{V - W^{||}}{V_{rel}} \qquad (5)$$

$$F_T^{\perp} = 0.24 V_{rel}^2 \sin\alpha = 0.24 V_{rel}^2 \frac{W^{\perp}}{V_{rel}} \qquad (6)$$

where V and W are the velocities of the cyclist and the wind with respect to the road.

The rider's power is now calculated as:

$$P = -\mathbf{F_T} \cdot \mathbf{V} = \left(F_R + F_D^{||}\right)V = 3.2V + 0.24 V_{rel}^2 \frac{V - W^{||}}{V_{rel}}V \qquad (7)$$

whereas the perpendicular force is simply balanced with the inclination of the cyclist.

To show the results of this model, we shall consider a cyclist rolling alone at 35 kmh[-1] (9.72 ms[-1]) on a flat road. This corresponds to the speed that a well-trained amateur cyclist can develop during long periods of time in the absence of wind. According to equation (7), for $W = 0$, the required power is about 250 W, an amount that the present authors have verified personally with the help of a Polar Power Output sensor coupled to a Polar S725i heart rate monitor [Polar, 2011].

It corresponds to a very ingenious system that works by measuring the tension and the speed of the transmission chain. The first sensor calculates chain tension by determining the frequency of vibration as it passes over a magnetic sensor. It suffices to enter the chain weight per unit length and the chain length. Another sensor positioned on the rear pulley also works magnetically and measures chain speed. The power is readily calculated as the product of both quantities. Assuming that our cyclist is pedaling at a constant power of 250 W under a wind speed of 25 kmh[-1] (6.94 ms[-1]), we summarize in table 1 the results for a rectilinear round-trip of 20 km considering no wind, a headwind, a tailwind or a sidewind. As expected, the duration of the trip increases with a headwind more than it decreases with a tailwind; over 20 km our cyclist loses 5 min and 33 s. But the most relevant result is the unexpected delay when cycling in a sidewind; over 20 km our cyclist increases his time by almost 3 min. Indeed, wind has a noteworthy effect on cycling [Íñiguez-de-la-Torre & Íñiguez, 2006].

Wind speed $W = 6.944$ m/s	Equation (7) (250 W constant power)	Cyclist speed V	Round-trip duration
No wind	$250 = 3.2V + 0.24V^3$	9.722 m/s (35 km/h)	34 min 17 s
Headwind	$250 = 3.2V + 0.24(V + 6.94)^2 V$	5.870 m/s (21.13 km/h)	39 min 50 s
Tailwind	$250 = 3.2V + 0.24(V - 6.94)^2 V$	14.576 m/s (52.47 km/h)	
Sidewind	$250 = 3.2V + 0.24(V^2 + 6.94^2)\dfrac{V}{\sqrt{V^2 + 6.94^2}}V$	9.006 m/s (32.42 km/h)	37 min 1 s

Table 2. Summary of the results for a 250 W cyclist under different wind conditions

With the purpose of describing the remarkable effect of an increasing lateral wind, the cyclist speed versus the sidewind speed for different power values is shown in figure 8. It paradoxical means that the crosswind leads to an increase in time. Intuitively, anyone would say that the lateral wind must be neutral. But all the cyclists well know its effect and recognize that this effect increases with their speed. Whereas an object at rest, $V \rightarrow 0$, would not be affected by this wind; when moving it will undergo an additional drag resistance due to the sidewind. To demonstrate this behavior, we shall neglect the rolling resistance in equation (5). For a sidewind of speed W, this is written as:

$$F_D^{|} = 0.24\left(V^2 + W^2\right)\cos\alpha = 0.24\ V\sqrt{V^2 + W^2} \tag{8}$$

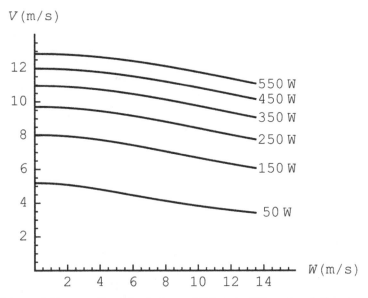

Fig. 8. Cyclist speed V versus the sidewind speed W, at six different pedaling power rates

clearly showing that for $V\rightarrow0$ the drag force is null, regardless of W. When V is non-zero, this aerodynamic drag force is always opposite to the direction of motion and it increases with V and W because

$$V\sqrt{V^2+W^2} \geq V^2 \qquad (9)$$

which shows that these results are a simple consequence of the quadratic dependence of the aerodynamic drag force on the air speed. In order to clearly understand this statement, it should be noted that if the air speed dependence were linear, equation (8) would be written as:

$$F_D^{||} = K\sqrt{V^2+W^2}\cos\alpha = KV \qquad (10)$$

(K representing the coefficient of proportionality for the hypothetically linear behavior of the aerodynamic drag force versus the air speed) acting as if there was no sidewind.

6. Cycling in team

When birds fly in formation use less energy to migrate that when they do it alone. Birds in flocks can therefore fly for more distance than when travelling on their own (see a flock of birds in V-formation in figure 9). Recently, *The Observer* published an article entitled "Geese point the way to saving jet fuel: Planes flying in V-formation are more efficient and produce less carbon dioxide", in such a manner that today the airline industry is looking to learn from them. In the future, our sky may perhaps not only have geese flying in V-formation but also passenger planes flying in group using less fuel and therefore producing less carbon dioxide emissions [McKie, 2009].

Fig. 9. Migration of a flock of geese V-formation

In 1914, a German researcher, Carl Wieselsberger, published a paper in which he showed that birds flying in V-formations use less energy to flap their wings than those on solo flights. He pointed out that when a bird flaps its wings it creates a current known as upwash; essentially, air lifts up and rises round the tips of the wings as they flap. Other birds, flying in the first one's wake, experience an updraft, allowing them to fly further. It is supported also by observations by French scientists who studied great white pelicans trained to fly behind an aircraft. The team, from the *Centre National de la Recherche Scientifique*, strapped instruments and transmitters to individual birds. This study revealed that the heart rate of the birds went down when they were flying together, and it also showed that they were able to glide more often when they flew in formation. Such experiments suggest that 25 large birds, such as pelicans or geese, flying in V-formation can travel 70 % further than flying alone. Many of the great migratory journeys, some covering thousands of miles, made by birds would be impossible without the energy-saving effects of group flight. Aviation engineers have now taken these discoveries to their logical conclusion and have proposed that aircrafts could fly in V-shaped groups so they can benefit from similar energy-saving effects.

This idea is the brainchild of researchers led by Professor Ilan Kroo, of Stanford University, California. In one calculation, the team envisaged three passenger jets leaving Los Angeles, Las Vegas and San Francisco airports en route to the east coast of the US. In the hypothetical exercise, the planes rendezvoused over Utah and then continued their journeys travelling in a V-formation, with planes taking turns to lead the formation. The group found that the aircrafts used 15 % less fuel and produced less carbon dioxide when flying in formation compared with solo performances. Such an approach could make significant inroads into the amount of carbon dioxide that is pumped into the atmosphere by planes.

However, critics have pointed to problems. Safety could be compromised by craft flying in tight formation, while coordinating departure times and schedules could become a major annoyance. The precision needed to coordinate just one aircraft to reach its destination safely is difficult enough, and air traffic controllers would be given a considerably larger headache if they had to coordinate the safe arrival several airplanes together. It also remains unproven as to how different weather conditions may affect V-formation flight efficiency. Although enabling flocks of passenger planes to fly in harmony like geese remain an arguably unlikely scenario, Chris Whitehead, a former hang glider pilot and teacher of the sport, remains skeptical of the California-based research. The local record holder and ex-Peak District team captain told that he specifically remembers how seagulls benefited by flying behind his glider, but other pilots coming in close proximity to one another would experience horrendous turbulence. It seems obvious that such difficulties could be only overcome with the help of more detailed research on their scheme.

As is well-known humans also imitate nature in this area; in many sports and motor competitions we apply the same aerodynamic and physic principles the birds use. The sport of cycling constitutes a beautiful example of how the members of a team can be helped to each other (see figure 10).

Fig. 10. Team time-trial in the *Tour de France*

In order to emphasize this issue, a brief experimental and computational study on the aerodynamics of a cycling team in a time-trial race is presented here. Special attention is paid to calculations of the power that each cyclist saves depending on his place in the team, and the results are compared with data from the individual time-trial (ITT) and the team time-trial (TTT) stages of the *Tour de France* for years of 2004 and 2009. In both cases, when comparing the racing of a rider alone with respect to that corresponding to the whole team, speed increments of up to 10 % and more are easily reached. There is a very few books in the literature addressing the field of aerodynamics in cycling, although on the web some sites devoted to analytical calculations in this area can be seen. Some recent works concerning aerodynamics and the basic physic principles of the sport of cycling can also be found [Wilson, 2004; di Prampero et al., 1979; Hannas & Goff 2004; Hannas & Goff, 2005; Landau & Lifshitz, 1982; Burke, 2003]. More details can be found in WebPages like [Bicycles Aerodynamics, 2011; Science of Cycling, 2011]

It is very difficult to perform analytical calculations of the power developed by cyclists when they are travelling in a group. This is only possible for a single rider. In this case, the total power (rolling, F_R, and aerodynamics, F_D) along a horizontal road with speed V, in the absence of wind is:

$$P = -\mathbf{F_T} \cdot \mathbf{V} = \left(F_R + F_D\right)V \qquad (11)$$

This equation can be written as:

$$P = \mu g (M + m) V + \frac{1}{2} \rho C_D A V^3 = \beta V + \gamma V^3 \tag{12}$$

where $\mu \leq 0.004$ is the rolling friction coefficient. While β takes values around 2.6 to 3.2 N, the γ coefficient takes values from 0.18 kgm^{-1}, in high competition cycling (for a medium-sized cyclist wearing high performance clothing and riding a time-trial bicycle), up to 0.22 kgm^{-1} for an amateur cyclist rolling in a semi-inclined position on the bicycle. Equation (11) shows good agreement with our experimental results as can be seen in figure 11.

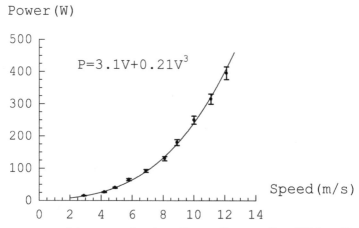

Fig. 11. Measurements of the power developed by a solitary cyclist of 79 kg rolling along a good horizontal road in the absence of wind. Data were obtained with the help of a Polar Power Output Sensor. Curve corresponds to equation (12) with $\beta \approx 3.1$ N and $\gamma \approx 0.21$ kgm^{-1}

Our power measurements in a small cycling group were made using the Conconi test (quasi-linear relation pulse versus power in the aerobic region) and a heart rate monitor and were verified in situ with the help of a Polar Power Output Sensor [Polar, 2011].

Powers of several hundred watt at speeds of 10–15 ms^{-1} clearly justify the importance of developing refined team strategies with a view to saving energy: rolling in pace line, forming compact groups (the so-called peloton) or arranging the riders diagonally when lateral wind blows (the well-known echelons) are the most frequent.

When cyclists travel in a group, the value of the β coefficient in equation (12) is strongly dependent on the geometry of the team (number and size of riders and gaps between them), and it also takes different values for each rider depending on his place in the group. Therefore, only individual power measurements would be able to provide precise information about the amount of energy that each cyclist uses up. In order to extend our experimental results to a larger group of cyclists, we resorted to performing a numerical simulation of the aerodynamics of cycling in a virtual wind tunnel.

Although a realistic analysis of the complex geometry of a cyclist would require using 3D fluid dynamics numerical simulation, the difficulties in its modeling counsel us to try a simpler analysis by means of a 2D numerical simulation. Therefore, we have used a software of a 2D virtual wind tunnel and a simplified model of the cyclists, just consisting of elliptic shapes that we have optimized to reproduce within 5 % our experimental results obtained for groups composed of five riders rolling in pace line with a gap between them of 20 cm. In this way, we were able to obtain good approximations of the aerodynamic characteristics (air pressure, temperature, density and wind speed) of larger groups.

The wind plot for a single cyclist rolling alone is in figure 12, while the case of a team composed of three cyclists is shown in figure 13. The power developed by the cyclists based on their position in the group is seen in table 3 [Íñiguez-de-la-Torre & Íñiguez, 2009]. The software here used, MicroCFD, is a Windows-based program on computational fluid dynamics for analyzing 2D planar and axisymmetric problems. Visualization is provided through color maps of local speed, density, pressure, and temperature [MicroCFD, 2011].

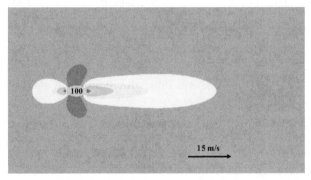

Fig. 12. Top-view of a simplified model of a 2D computational air-speed flow picture for a single cyclist rolling alone in a virtual wind tunnel at 15 ms^{-1}. The aerodynamic drag force is normalized to F = 100. Warm/cool colors represent high/low air-speeds

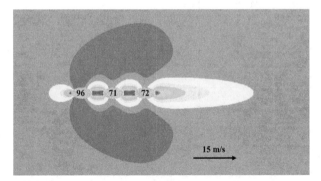

Fig. 13. Top-view of a computational air-speed flow picture for a team composed of three cyclists when they travel in a virtual wind tunnel at 15 ms^{-1} in pace line. The aerodynamic drag forces are respectively 96, 71 and 72 % of that corresponding to a single rider rolling alone. The purple (darkest) color corresponds to the artificial tailwind

As anticipated, it may be observed that the cyclists who travel behind have to make less effort but it should be noted that the cyclist in the front also benefits from the presence of the rest of the team; the artificial tailwind also helps him by about 5 %. The aerodynamic coefficient is effectively reduced. This result, unexpected for many people, is known, or at least suspected, by cyclists accustomed to rolling in groups at high speeds because all the effects are enhanced as speed increases.

As expected, the advantage of cycling in a well-coordinated group increased rapidly with the number of components, and our study showed that the ideal number of a team is five or six cyclists and that the mean value of the force diminishes to 70 % of that corresponding to a single cyclist rolling alone. Indeed, when the size of the group grows at the time of taking turns as leader (which reduces the effective number of riders), the difficulties in coordination increase, and correctly preserving the gap between the bicycle wheels becomes difficult. Certain other complications involved in the difficulties of maintaining the pace line are the presence of curves, sidewind or the different size and power of the cyclists. Finally, it should be noted that it is not easy to continuously preserve such a small separation between cyclists (10–20 cm) when rolling at speeds so high as 10–15 ms^{-1}.

In the TTT of the *Tour de France* 2004, the team of Lance Armstrong travelled at 14.92 ms^{-1}, whereas in the ITT he rolled at an average speed of 13.72 ms^{-1}, over similar roads in the length and profile. More recently, in the TTT of the 2009 edition of the *Tour*, the average speed of the Astana team was 13.98 ms^{-1}, although unfortunately the characteristics of the stage were not appropriate for the practice of cycling in a well-synchronized team. In contrast, the speed of Alberto Contador in the ITT was 13.75 ms^{-1}, practically identical to the speed of Armstrong in the 2004 ITT. Upon comparing these results with our calculations, it may be concluded that travelling in a group in a real stage is not as advantageous as the calculation predicts, doubtless due to the reasons alleged in the previous paragraph.

Team size	F_1	F_2	F_3	F_4	F_5	F_6	F_7	F_8	F_9	<F>
1	100									100
2	96	74								85
3	96	71	72							80
4	96	69	68	70						76
5	95	68	67	67	69					73
6	95	68	67	67	67	69				72
7	95	67	66	66	66	66	68			71
8	95	67	66	66	66	66	66	68		70
9	95	67	66	66	66	66	66	66	68	70

Table 3. Forces that each cyclist develops when travelling in pace line as a function of his place in the group. The values are normalized (F = 100) to that corresponding to a cyclist rolling alone. The last column shows the average value of the force developed by the riders of the team as a function of the group size.

7. Conclusions

It has been demonstrated how the study of the aerodynamics in low speed vehicles is especially interesting when the available power is scarce. For that purpose, the analysis of the aerodynamics of cycling has been revealed like a fantastic example that has allowed showing how the investigation in cycling equipment, fairing, position and adequate clothes of the rider is essential.

The detailed analysis of the not well-known sidewind effects on the cycling race and the study of the saved power by each rider when cycling in compact group have shown a remarkable interest. It has been pointed out the significant breaking effect of the lateral wind on the riders and how the artificial tailwind that a group creates at moderate speed, constitutes a powerful help in high competition races or in simple recreational trips like many groups of amateur cyclists around the world do every week-end.

8. References

Bent H. A. (1978). Energy and exercise I: How much work can a person do? Journal of Chemical Education 55, 456

Bicycles Aerodynamics. In: *Bicycles Aerodynamics*, October 2011, Available from:
< http://www.sheldonbrown.com/rinard/aero/index.htm >

Burke E. R. (2003). High Tech Cycling. The Science of Riding Faster. Human Kinetics, USA.

Chandler D. (2008) On Gossamer wings. Record-breaking Daedalus project marks 20th anniversary, TechTalk Serving the MIT Community, 52-24, April 30

Cleary P. A. and Mohazzabi P. (2011). On the stability of a bicycle on rollers, European Journal of Physics 32, 1293

Cyclingforums. In: *Tour De France: Power Outputs*, October 2011, Available from:
< http://www.cyclingforums.com/t/468821/tour-de-france-power-outputs >

di Prampero P. E., Cortili G., Mognoni P. and Saibene F. (1979). Equation of motion of a cyclist, J. Appl. Physiol. 47, 201

Eriksen H. K., Kristiansen J. R., Langangen Ø. and Wehus I. K. (2009). How fast could Usain Bolt have run? A dynamical study, American Journal of Physics, 77-3, 224

Gross A. C., Chester R. K. and Douglas J. M. (1983). The aerodynamics of land vehicles, Scientific American 249, no. 9

Hannas B. L. and Goff J. E. (2004). Model of the 2003 Tour de France, Am. J. Phys. 72 575-578

Hannas B. L. and Goff J. E. (2005). Inclined-plane model of the 2004 Tour de France, Eur. J. Phys. 26 251-259

HPVA. In: *The International Human Powered Vehicle Association*, October 2011, Available from:
< http://hpva.us/ >

Indexmundi. In: *World Crude Oil Consumption by Year*, October 2011, Available from:
< http://www.indexmundi.com/energy.aspx >

Íñiguez-de-la-Torre A. and Íñiguez J. I. (2009). Aerodynamics of a cycling team in a time trial: does the cyclist at the front benefit?, Eur. J. Phys. 30, 1365–1369

Íñiguez-de-la-Torre I. and Íñiguez J. I. (2006). Cycling and wind: does sidewind brake?, Eur. J. Phys. 27, 71–74

Jones D. E. H. (1970). The stability of the bicycle, Physics Today 23, 34 (reprinted in September 2006)

Kooijman J. D. G., Meijaard J. P., Papadopoulos J. M., Ruina A. and Schwab A. L. (2011). A bicycle can be self-stable without gyroscopic or caster effects, Science 332, 339

Landau L. D. and Lifshitz E. M. (1982) Fluid Mechanics 2nd edn, Oxford: Pergamon

Lea & Febiger. (1991). Guidelines for Exercise Testing and Prescription, American College of Sports Medicine, 4th edition, Philadelphia

McKie R. (2009). Science Editor The Observer, Sunday 27 December

MicroCFD. In: *MicroCFD*, October 2011, Available from:
< http://www.microcfd.com/ >

Padilla S., Múgica I., Angulo F. and Goiriena J. J. (2000). Scientific approach to the 1 h cycling world record: A case study, Journal of Applied Physiology, 89, 1522

Polar. In: *Polar*, October 2011, Available from:
< http://www.polar.fi/en/ >

Rotorbike. In: *Rotorbike*, October 2011, Available from:
< http://www.rotorbike.com/ >

Science of Cycling. In: *Science of Cycling*, October 2011, Available from:
< http://www.exploratorium.edu/cycling/index.html >

Sonawat H. M., Phadke R. S. and Govil G. (1984) Towards biochemical fuel cells, Proc. Indian Acad. Sci. (Chem. Sci.), 93-6, 1099

Wikipedia. In: *Daedalus (avion)*, October 2011, Available from:
< http://fr.wikipedia.org/wiki/Daedalus_(avion) >

Wilson D. G. (1997). Wind-tunnel tests: Review of Tour, das Radmagazin, article, Human power 12, no 4

Wilson D. G. (2004). Bicycling Science, 4th edition, MIT Press, Cambridge, MA

Section 2

Aerodynamics and Flow Control

3

Bluff Body Aerodynamics and Wake Control

Efstathios Konstantinidis and Demetri Bouris
Department of Mechanical Engineering
University of Western Macedonia
Bakola & Sialvera, Kozani
Greece

1. Introduction

In aerodynamics, a bluff body is one which has a length in the flow direction close or equal to that perpendicular to the flow direction. This spawns the characteristic that the contribution of skin friction is much lower than that of pressure to the integrated force acting on the body. Even a streamlined body such as an airfoil behaves much like a bluff body at large angles of incidence. A circular cylinder is a paradigm often employed for studying bluff body flows.

Bluff body flows are characterized by flow separation which produces a region of disturbed flow behind, i.e. the wake. Within the near-wake various forms of flow instabilities, both convective and absolute, may be triggered and amplified. These instabilities are manifested by the generation of two- and three-dimensional unsteady flow structures and eventually turbulence as the Reynolds number ($Re = U_\infty D / \nu$, where U_∞ is the incident flow velocity, D is the width of the body and ν is the kinematic viscosity of the fluid) is progressively increased. The most well-known instability is that leading to the periodic formation and shedding of spanwise vortices which produce an impressive wake pattern named after von Kármán and Benhard. The frequency of vortex shedding is characterized by the Strouhal number,

$$St = \frac{f_{vs}D}{U_\infty} \tag{1}$$

which is a function of the Reynolds number, $St(Re)$. The self-sustained wake oscillations · associated with vortex shedding are incited through a Hopf bifurcation as predicted from linearized stability theory (Triantafyllou et al., 1987) and remain in tact over a wide range of Reynolds numbers from approximately 50 to 10^6 or even higher.

Bluff body flows involve the interaction of three shear layers, namely the boundary layer, the separating free shear layers and the wake. The physics of vortex formation and the near-wake flow have been the subject of many past experimental and numerical studies which have provided a wealth of information. The reader is referred to Bearman (1997) for a pertinent review. The basic mechanism of vortex formation is essentially two dimensional although there are inherent three-dimensional features for Reynolds numbers above 190. As well as contributing to time-averaged aerodynamic forces, vortex shedding is responsible for the generation of fluctuating forces acting on the body. This has several ramifications in practical applications as it may excite flow-induced vibration and acoustic noise.

In order to modify the aerodynamic characteristics of bluff bodies, e.g., for control of drag, vibration and noise, it is imperative to control the separated flow in the wake and the dynamics of vortex formation which are the sources of the fluid forces on the body. Such control can be achieved by both active and passive methods. In active control, it is typical to perturb the wake flow by some sort of excitation mechanism, e.g., rectilinear or rotational oscillations of the body, flow and/or sound forcing, fluid injection and/or blowing (synthetic jets), etc. Some of these methods affect the local flow while others influence the flow globally. The frequency and amplitude of the imposed perturbations provide a means to achieve the desired objective. Passive control is achieved by modification of the cylinder's geometry, e.g., by the addition of tabs, strakes or bumps; these mainly influence the three-dimensional flow characteristics.

A main characteristic of the wake response to periodic perturbations is the 'vortex lock-on' phenomenon whereby the imposed perturbation takes over control of the vortex formation and shedding over a range of conditions in the amplitude–frequency parameter space. For excitation along the flow direction, lock-on occurs when the driving frequency is near twice that of vortex shedding whereas the perturbations need be near the vortex shedding frequency for perturbations transverse to the flow direction. Lock-on is accompanied by a resonant amplification of both mean and fluctuating forces acting on the body but other types of wake response can take place, e.g., different modes of vortex formation can be incited such as symmetric (Konstantinidis & Balabani, 2007). These interactions provide a means for wake flow control.

This chapter presents issues related to the fluid dynamics of bluff bodies in steady and time-dependent flows. The importance of vortex shedding for the generation of aerodynamic forces is exemplified with reference to numerical simulations of two-dimensional fluid flow about a circular cylinder at low Reynolds numbers. Results are presented for the flow patterns in the near-wake and fluid-induced forces exerted on the cylinder in response to flows with superimposed harmonic and non-harmonic perturbations in velocity. Implications for wake control are also discussed. It seems reasonable to concentrate on the stable range of Reynolds number (laminar wake regime) where the effects of simple changes can be studied without the complications attending the presence of turbulent flow.

2. Methodology

In order to study the fluid flow about a circular cylinder and derive the aerodynamic loading, the Navier–Stokes (N–S) equations are solved numerically using a finite volume method on an orthogonal curvilinear grid in two dimensions (2D). The orthogonal curvilinear grid is an appropriate choice for the description of the flow geometry because it can be readily adapted to the cylinder surface while reducing numerical diffusion without introducing overly complex terms in the equations as is the case of general curvilinear or unstructured grids. The governing equations can be written as

$$\frac{\partial}{\partial t}(\rho\Phi) + \frac{1}{l_\xi l_\eta}\frac{\partial}{\partial\xi}(\rho u l_\eta \Phi) + \frac{1}{l_\xi l_\eta}\frac{\partial}{\partial\eta}(\rho v l_\xi \Phi) = \\ \frac{1}{l_\xi l_\eta}\frac{\partial}{\partial\xi}(\mu \frac{l_\eta}{l_\xi}\frac{\partial\Phi}{\partial\xi}) + \frac{1}{l_\xi l_\eta}\frac{\partial}{\partial\eta}(\mu \frac{l_\xi}{l_\eta}\frac{\partial\Phi}{\partial\eta}) + S_\Phi \tag{2}$$

where ρ and μ is the density and viscosity of the fluid, respectively, l_ξ and l_η are the spatially varying metric coefficients related to the orthogonal curvilinear coordinates (ξ, η), and (u, v) the corresponding velocities in local coordinates. The variable $\Phi = 1, u, v$ for the continuity and momentum equations, respectively, and S_Φ are the source terms, including pressure terms. The source term is zero for the continuity equation, whereas

$$S_u = -\frac{1}{l_\xi}\frac{\partial p}{\partial \xi} - \frac{\rho uv}{l_\xi l_\eta}\frac{\partial l_\xi}{\partial \eta} + \frac{\rho v^2}{l_\xi l_\eta}\frac{\partial l_\eta}{\partial \xi} + \frac{1}{l_\xi l_\eta}\frac{\partial}{\partial \xi}\left[\frac{l_\eta}{l_\xi}\mu\left(\frac{\partial u}{\partial \xi} + \frac{2v}{l_\eta}\frac{\partial l_\xi}{\partial \eta}\right)\right] +$$

$$\frac{1}{l_\xi l_\eta}\frac{\partial}{\partial \eta}\left[l_\xi\mu\left(\frac{1}{l_\xi}\frac{\partial v}{\partial \xi} - \frac{v}{l_\xi l_\eta}\frac{\partial l_\eta}{\partial \xi} - \frac{u}{l_\xi l_\eta}\frac{\partial l_\xi}{\partial \eta}\right)\right] + \qquad (3)$$

$$\frac{\mu}{l_\xi l_\eta}\left(\frac{1}{l_\xi}\frac{\partial v}{\partial \xi} + \frac{1}{l_\eta}\frac{\partial u}{\partial \eta} - \frac{v}{l_\xi l_\eta}\frac{\partial l_\eta}{\partial \xi} - \frac{u}{l_\xi l_\eta}\frac{\partial l_\xi}{\partial \eta}\right)\frac{\partial l_\xi}{\partial \eta} - \frac{2\mu}{l_\xi l_\eta}\left(\frac{1}{l_\eta}\frac{\partial v}{\partial \eta} + \frac{u}{l_\xi l_\eta}\frac{\partial l_\eta}{\partial \xi}\right)\frac{\partial l_\eta}{\partial \xi}$$

for $\Phi = u$, and

$$S_v = -\frac{1}{l_\eta}\frac{\partial p}{\partial \eta} - \frac{\rho uv}{l_\xi l_\eta}\frac{\partial l_\eta}{\partial \xi} + \frac{\rho u^2}{l_\xi l_\eta}\frac{\partial l_\xi}{\partial \eta} + \frac{1}{l_\xi l_\eta}\frac{\partial}{\partial \eta}\left[\frac{l_\xi}{l_\eta}\mu\left(\frac{\partial v}{\partial \eta} + \frac{2u}{l_\xi}\frac{\partial l_\eta}{\partial \xi}\right)\right] +$$

$$\frac{1}{l_\xi l_\eta}\frac{\partial}{\partial \xi}\left[l_\eta\mu\left(\frac{1}{l_\eta}\frac{\partial u}{\partial \eta} - \frac{v}{l_\xi l_\eta}\frac{\partial l_\eta}{\partial \xi} - \frac{u}{l_\xi l_\eta}\frac{\partial l_\xi}{\partial \eta}\right)\right] + \qquad (4)$$

$$\frac{\mu}{l_\xi l_\eta}\left(\frac{1}{l_\xi}\frac{\partial v}{\partial \xi} + \frac{1}{l_\eta}\frac{\partial u}{\partial \eta} - \frac{v}{l_\xi l_\eta}\frac{\partial l_\eta}{\partial \xi} - \frac{u}{l_\xi l_\eta}\frac{\partial l_\xi}{\partial \eta}\right)\frac{\partial l_\eta}{\partial \xi} - \frac{2\mu}{l_\xi l_\eta}\left(\frac{1}{l_\xi}\frac{\partial u}{\partial \xi} + \frac{v}{l_\xi l_\eta}\frac{\partial l_\xi}{\partial \eta}\right)\frac{\partial l_\xi}{\partial \eta}$$

for $\Phi = v$. Volume averaging is performed in physical space and the metric coefficients are replaced by the physical distances. All variables are collocated at the grid cell centers so the original SIMPLE algorithm is combined with the Rhie and Chow modification to avoid checker board oscillations in the pressure coupling (Rhie & Chow, 1983). A fully-implicit first-order Euler discretization of the temporal term is used while for the convection and diffusion terms, a higher-order bounded upwind scheme (Papadakis & Bergeles, 1995) and central differencing is employed, respectively.

The grid is initially constructed for half the cylinder and then symmetrically duplicated so that the solid cylinder body is comprised of a number of grid cells defined by two subsequent grid lines in the ζ direction and a number of grid lines in the η direction. Velocity and pressure are artificially set to zero at these cells and they are treated as solid boundaries (no-slip condition) by all neighboring cells. At the lower and upper boundaries of the domain, symmetry conditions are employed. A steady or time-dependent velocity specified at the inlet of the flow domain whereas a convective boundary condition is employed at the outlet in order to avoid backward reflection of pressure waves. The computational domain is rectangular and extends $10D$ upstream, $25D$ downstream and $10D$ above and below the cylinder. The orthogonal curvilinear mesh consists of 299×208 nodes, which is sufficient for mesh-independent results. A time step of the order of $\delta t U_\infty / D = 10^{-2}$ is employed in the simulations. The system of discretized equations is solved using a Tri-Diagonal Matrix Algorithm (TDMA) with an iterative Alternating Direction Implicit (ADI) method.

The sectional forces acting on the cylinder are calculated from the integration of skin friction and pressure around its periphery as

$$F_X = \int_0^{2\pi} \tau_w \sin\theta d\theta + \int_0^{2\pi} p \cos\theta d\theta \qquad (5)$$

$$F_Y = \int_0^{2\pi} \tau_w \cos\theta d\theta - \int_0^{2\pi} p \sin\theta d\theta \qquad (6)$$

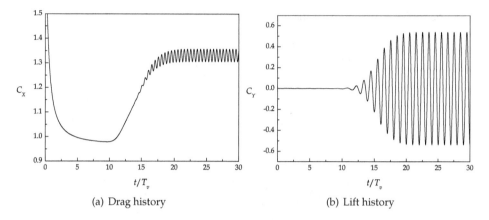

(a) Drag history

(b) Lift history

Fig. 1. Instantaneous in-line (drag) and transverse (lift) forces for $Re = 150$.

where θ is the angle measured clockwise from the front stagnation point. The force is nondimensionalized to yield the in-line (drag) and transverse (lift) force coefficients according to formulae

$$C_X = \frac{F_X}{\frac{1}{2}\rho U_\infty^2 D} \quad \text{and} \quad C_Y = \frac{F_Y}{\frac{1}{2}\rho U_\infty^2 D} \tag{7}$$

3. Circular cylinder in steady flow

In this section, results are shown for steady incident flow. The evolution of the instantaneous drag and lift coefficients is shown in Fig. 1 for a typical run at $Re = 150$. In this simulation, the velocity was set to zero everywhere in the flow domain. It takes more than ten periods of vortex shedding for the flow oscillations to be initiated in the wake. During this period, the instantaneous drag drops rapidly from very high values, then reaches a plateau before rising again in Fig. 1(a). Once the instability is triggered, the oscillations are amplified and become saturated in approximately another ten periods as it is seen more clearly in the history of the lift in Fig. 1(b). Beyond this point, a steady-state periodic oscillation is established.

Figure 2 shows the vorticity distribution around the cylinder at six instants over approximately half vortex shedding cycle in the steady periodic state. The instantaneous forces on the cylinder are also shown at the bottom subplots and symbols (red circles) indicate the different instants shown. The figure shows the processes of vortex formation and shedding in the near wake. Vortices are regions of low pressure and as they are formed near the cylinder, a fluctuating pressure field acts on the cylinder surface. As a result, the drag is minimized while a positive vortex (marked by red contour lines) is formed on the bottom side of the cylinder in phases 1 and 2. Subsequently, this positive vortex moves towards the opposite side and away from the cylinder and the drag increases to its maximum in phase 4. In phases 5 and 6, a negative vortex begins to form on the top side of the cylinder which, in turn, will induce a low back pressure and a minimum in drag as it gains strength and size. A maximum in the lift is observed between phases 3 and 4 which is associated with the entrainment of the fluid at the end of the formation region across the wake centreline and towards the top side which causes the negative vortex on the same side to be shed. This process repeats periodically and a staggered vortex street is formed downstream in the wake.

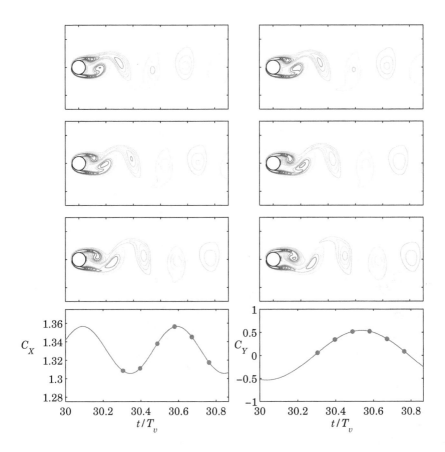

Fig. 2. Vorticity distributions in the cylinder wake and fluid forces on the cylinder at different instants in steady flow; $Re = 150$.

The drag fluctuates at a frequency twice that of the lift since each vortex shed from either side of the cylinder induces a cyclic oscillation in the in-line direction. Hence, the frequency at which vortices are shed from the one side of the cylinder is equal to the frequency of the lift. To some degree of approximation, the instantaneous forces can be represented by harmonic functions

$$C_X(t) = C_D + \sqrt{2}C_D' \sin\left(4\pi f_v t + \phi_D\right), \tag{8}$$

$$C_Y = \sqrt{2}C_L' \sin\left(2\pi f_v t + \phi_L\right). \tag{9}$$

Figure 3 shows some time-averaged statistics of the velocity and pressure fields at $Re = 150$. Velocity is normalized with the incident flow velocity, U_∞, and pressure is normalized with $\frac{1}{2}\rho U_\infty^2$. The distribution of the mean streamwise velocity shows the acceleration of the fluid as it passes over the 'shoulders' of the cylinder and the 'recirculation bubble' (reverse velocities) behind it (Fig. 3(a)). The bubble closure point where the mean velocity becomes zero is located 1.6 diameters behind the cylinder centre. When Bernoulli's equation is applied to the flow

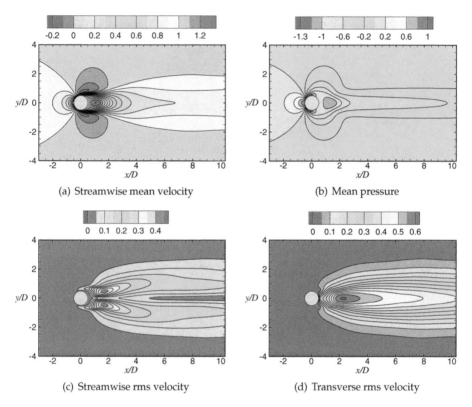

(a) Streamwise mean velocity

(b) Mean pressure

(c) Streamwise rms velocity

(d) Transverse rms velocity

Fig. 3. Time-averaged flow characteristics for $Re = 150$.

outside separation, the back pressure coefficient is given by (Griffin & Hall, 1991)

$$C_{pb} = \frac{p_b - p_\infty}{\frac{1}{2}\rho U_\infty^2} = 1 - K^2 \tag{10}$$

where $K = U_s/U_\infty$ is the velocity ratio at separation. The maximum velocity magnitude outside separation is $1.36U_\infty$ at a distance $0.23D$ from the surface. Figure 3(b) shows the distribution of the mean pressure where a back pressure coefficient of $C_{pb} = -0.95$ is obtained, compared to a value of -0.85 computed from the above equation. The mean pressure coefficient attains a minimum value of -1.36 on the surface of the cylinder at $\theta = 85°$ which, coincidentally, is exactly equal to the velocity ratio slightly further from the surface. The point of time-averaged flow separation is located at $\theta_s = 112°$. This compares well to the value of $114°$ computed from the following empirical relationship

$$\theta_s = 101.5 + 155.2Re^{-1/2}. \tag{11}$$

The above equation was suggested by Wu et al. (2004) who conducted soap-film experiments and numerical simulations under 2D flow conditions.

	Present study			Henderson (1995)		
Re	C_{Dv}	C_{Dp}	C_D	C_{Dv}	C_{Dp}	C_D
60	0.444	0.985	1.429	0.432	0.984	1.415
90	0.368	0.996	1.364	0.362	0.998	1.360
120	0.323	1.017	1.340	0.319	1.020	1.339
150	0.292	1.040	1.331	0.289	1.044	1.333
180	0.269	1.064	1.334	0.267	1.069	1.336

Table 1. Viscous, pressure and total drag for different Reynolds numbers.

The vortex shedding produces an unsteady flow in the wake that can be characterized by the root-mean-square (rms) magnitude of the velocity fluctuations. Figure 3(c) shows that the streamwise rms velocity exhibits a double-peaked distribution reaching a maximum value of $0.42U_\infty$ at $\{x/D, y/D\} = \{1.67, \pm0.47\}$. The transverse rms velocity exhibits a single-peaked distribution with a maximum value of $0.60U_\infty$ at $x/D = 2.22$ on the wake centreline as shown in Fig. 3(d). The maxima in the rms velocities are approximately 5% higher than those measured in the cylinder wake at $Re = 2150$ (Konstantinidis & Balabani, 2008). This observation suggests that the mechanism of vortex formation remains basically unaltered over this Re range so long as the boundary layer on the cylinder surface remains laminar until separation.

Simulations were carried out at different Reynolds numbers in the range $Re = 60 - 180$ and the global flow characteristics were determined. One of the key parameters in cylinder wakes is the mean drag coefficient, C_D, which is indicative of the mean energy dissipated in the fluid. Table 1 shows the variation of the mean drag coefficient. The table includes the contribution of the skin friction C_{Dv} and pressure C_{Dp} components to the total force. The skin friction drag reduces monotonically but non-linearly with increasing Reynolds number as might be expected. On the other hand, the pressure drag increases with increasing Reynolds number. As a result, the total drag exhibits a minimum at $Re = 150$. Results are also shown from another numerical study using a high-order spectral element code (Henderson, 1995). The maximum difference of the total drag between the present study and that of Henderson is 1%.

Figure 4 shows the variation of the fluctuating lift coefficient and the Strouhal number as a function of the Reynolds number. Both variables increase as the Reynolds number increases. The solid line in Fig. 4(a) is an empirical fit to experimental data for the sectional lift coefficient provided in Norberg (2003),

$$C'_L = \left(\frac{\epsilon}{30} + \frac{\epsilon^2}{90} \right)^{1/2} \tag{12}$$

where $\epsilon = (Re - 47)/47$. The above equation is valid in the range $Re = 47 - 190$. The present simulations predict a lift coefficient which is slightly above the line representing the empirical relationship. Although small, this difference can be attributed to three-dimensional effects inevitably present in experiments and, in particular, the initiation of slantwise vortex shedding (Gerrard, 1966). Norberg (2003) also provided an empirical relationship for the Strouhal number in the same range $Re = 47 - 190$,

$$St = 0.2663 - 1.019\sqrt{Re}. \tag{13}$$

It should be pointed out that the Strouhal number refers to the frequency at which vortices are shed from one side of the cylinder. The present results agree fairly well with the empirical relationship even though it is based on experimental data.

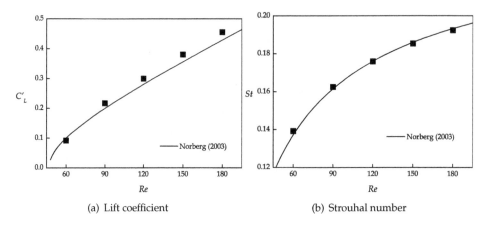

(a) Lift coefficient   (b) Strouhal number

Fig. 4. Flow parameters as a function of Reynolds numbers in steady flow. Symbols show the present results while the lines indicate empirical relationships.

Overall, the results presented here for a range of Reynolds numbers in the laminar unsteady wake regime for steady incident flow compare favourably with experimental data and other numerical simulations which provides confidence in the numerical method employed. The simultaneous presentation of the vortex patterns in the wake and fluid forces on the body has revealed effectively the connection between them.

4. Circular cylinder in harmonically perturbed flow

The wake response to harmonic perturbations in the velocity of the incident flow is examined in this section. In this case, two more dimensionless parameters number are required to describe the flow in addition to the Reynolds. The time-depended incident flow velocity is given by

$$U(t) = U_\infty + \Delta u \sin{(4\pi f_e t)} \qquad (14)$$

where Δu is the amplitude and f_e is the nominal frequency of the imposed velocity perturbations. It should be pointed out that the actual frequency of the velocity waveform in Eq. (14) is $2f_e$ for consistency with non-harmonic waveforms that will be employed in the next section. In the following, the different cases examined will be described in terms of

$$\text{Reduced velocity} = \frac{U_\infty}{f_e D},$$

$$\text{Amplitude ratio} = \frac{\Delta u}{U_\infty}.$$

Results are shown below for a constant Reynolds number ($Re = 150$) and a constant amplitude ratio ($\Delta u/U_\infty = 0.20$). The interest here is in the determination of the vortex lock-on range in which the shedding frequency is captured by the perturbation frequency. For this purpose, a number of runs were conducted starting from perturbation frequencies near the vortex shedding frequency in the unperturbed wake for which lock-on is guaranteed, and either increasing or decreasing the perturbation frequency until the wake response is not locked-on.

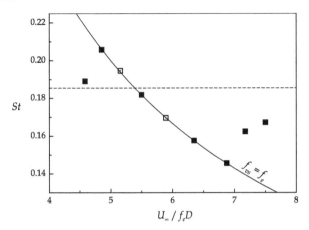

Fig. 5. Variation of the Strouhal number *vs.* reduced velocity.

Figure 5 shows the frequency of vortex shedding in terms of the Strouhal number, $St = f_{vs}D/U_\infty$, as a function of the excitation frequency in terms of the reduced velocity. If there was no interaction between the imposed perturbations and vortex shedding, the Strouhal number would remain constant at the value indicated by a horizontal dashed line. However, the Strouhal number is affected for all the perturbation frequencies considered. For a range of reduced velocities, vortex shedding locks on to the perturbation frequency as indicated by the solid line. For low reduced velocities, the Strouhal number gets higher than its unperturbed value in the natural wake, and vice versa for high reduced velocities. The two cases marked by open symbols will be examined in more detail below.

Figure 6 shows the instantaneous vorticity distributions for two different reduced velocities together with the variation of the drag and lift forces at three instants over approximately half vortex shedding cycle (cf. Fig. 2). For both cases, single vortices are shed alternatively from each side of the cylinder in a fashion similar to that observed for steady flow. However, the details of vortex formation are different. For example, the peak vorticity values and the size of the vortices is increased in harmonically perturbed flow due to wake resonance. Furthermore, the longitudinal distance between shed vortices varies linearly with the reduced velocity. These changes in the wake patterns are accompanied by considerable changes in the magnitude of the mean and fluctuating forces acting on the cylinder. The unsteady in-line force shows a fifty-fold increase in magnitude whereas the transverse increases by 2.5 times compared to their counterparts in steady flow. The in-line force in unsteady flow contains inertial aerodynamic components due to added mass and pressure gradients, i.e.,

$$F_M = \frac{1}{4}\rho\pi D^2 C_M \frac{dU}{dt} \tag{15}$$

where $C_M = 1 + C_A$ is the inertia coefficient and C_A the added mass coefficient. Under the conditions examined here, the inertial components of the in-line force dominate over the viscous component. As a consequence, the in-line force is almost in phase with the flow acceleration dU/dt. On the other hand, the transverse force is not affected by inertial components since there is no perturbation in that direction. In Fig. 6, the waveform of the transverse force deviates markedly from a pure harmonic for both reduced velocities and there is quite a difference between them reflecting the effect of perturbation frequency on C_Y.

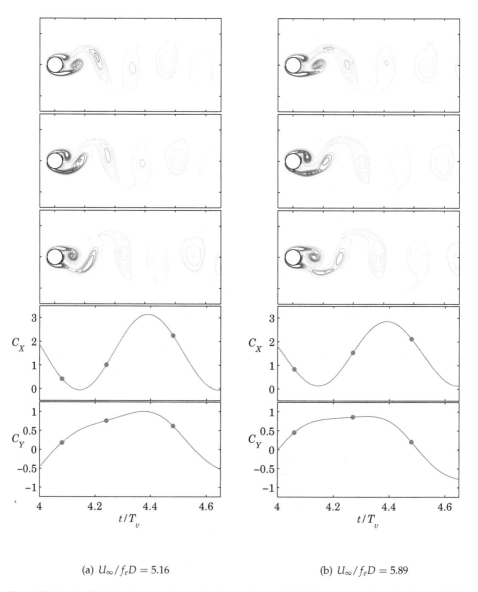

(a) $U_\infty/f_e D = 5.16$ (b) $U_\infty/f_e D = 5.89$

Fig. 6. Vorticity distributions in the cylinder wake and fluid forces on the cylinder at different instants for two reduced velocities in harmonically perturbed flow; $Re = 150$, $\Delta u/U_\infty = 0.20$.

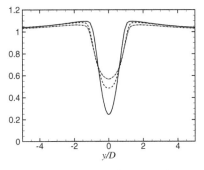

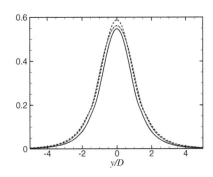

(a) Streamwise mean velocity at $x/D = 1.98$ (b) Transverse rms velocity at $x/D = 3.52$

Fig. 7. Time-averaged flow characteristics for steady and harmonically perturbed incident flow; Solid line: steady flow, dashed line: $U_\infty/f_eD = 5.89$, dash-dot line: $U_\infty/f_eD = 5.16$. All cases: $Re = 150$, $\Delta u/U_\infty = 0.20$.

For $U_\infty/f_eD = 5.89$, the lift attains a very broad maximum which makes it difficult to define precisely the phase, ϕ_L. In other words, the approximation of the lift force as a harmonic function is very crude. Nevertheless, it can be readily verified that the phase at which $C_Y(t)$ is maximum shifts with reduced velocity. E.g., it occurs at $t/T_v \approx 4.4$ for $U_\infty/f_eD = 5.16$ and at a slightly earlier time for $U_\infty/f_eD = 5.89$. In effect, the perturbation frequency provides a means to control the phasing of vortex shedding from the cylinder.

Different perturbation frequencies not only affect the unsteady flow characteristics but also the time-averaged flow. Figure 7 shows the distributions of mean and rms velocities across the wake for two different reduced velocities in harmonically perturbed flow compared to those in steady flow. Clearly, the distributions are much affected in perturbed flow but also depend on the reduced velocity. When looking at the time-averaged location of flow separation it is observed that $\theta_s = 119°$ and $117°$ for $U_\infty/f_eD = 5.16$ and 5.89, respectively. Hence, the separation point moves forward in perturbed flow compared to its location in steady flow.

The variation of the mean drag coefficient over the range of reduced velocities examined for $\Delta u/U_\infty = 0.20$ is shown in Fig. 8. The data show a drag amplification at a reduced velocity of 5.15 reaching a maximum value which is 16% higher than its value for steady flow. Outside lock-on response, the drag coefficient is close to that for the unperturbed flow.

5. Effect of non-harmonic flow perturbations

In this section, the effect of modest deviations of the perturbation waveform from a pure harmonic on the wake flow and fluid forces on the cylinder is examined. The time-depended incident flow velocity is now given by

$$U(t) = \beta \pm \left[1 + \alpha \sin^2(2\pi f_e t)\right]^n \tag{16}$$

where α is a parameter related to the amplitude of velocity perturbations, β sets the mean and median velocity, and the index n dictates the perturbation waveform. For $n = 1$, the waveform is pure-tone harmonic as those employed in the previous section. For $n \neq 1$, the perturbation waveform comprises an infinite number of harmonics. The effect of different values of n has

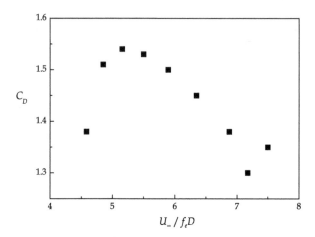

Fig. 8. Variation of the mean drag coefficient *vs.* reduced velocity.

been examined previously by the authors (Konstantinidis & Bouris, 2010). For the present study, a value of $n = -1$ was employed and two different waveforms were produced by using different signs in Eq. (16). Therefore, the waveform produced by a plus sign will be denoted 'plus waveform' and likewise for the 'minus waveform'. The alpha parameter was kept constant so that the velocity amplitude $\Delta U = (U_{max} - U_{min})/2 = 0.20U_\infty$, i.e., the same as that for harmonic perturbations. It should be noted that the time-averaged mean velocity of the incident flow is slightly different for the two different waveforms. Hence, the median velocity $U_0 = (U_{max} + U_{min})/2 \approx U_\infty$ is employed as a reference to allow comparisons. Furthermore, due to the large number of independent variables, results will be shown below only for the two perturbation frequencies which have been examined in more detail in the previous section to allow comparisons between different non-harmonic and harmonic waveforms.

Figure 9 shows the instantaneous vorticity distributions for the two different waveforms together with the drag and lift forces at three instants over approximately half vortex shedding cycle at a reduced velocity of 5.16 (cf. Fig. 2(a)). For both waveforms, the wake response remains phase-locked to the imposed perturbation. Comparison of the vorticity contour lines indicates that the details of vortex formation depend on the perturbation waveform. The plus waveform causes the formation of slightly larger vortices than the minus waveform. Furthermore, the vortices appear to be shed slightly earlier for the minus waveform whereas a plus waveform appears to delay vortex shedding. However, it is difficult to quantify these differences. On the other hand, their effect is more evident on the variation of the fluid forces. The in-line force is markedly different for the two waveforms as it follows the flow acceleration dU/dt due to the dominance of the inertial components. The waveform of the transverse force exhibits some delicate differences, most notably, the time at which it attains its maximum which occurs at a slightly later time for the plus than for the minus waveform. This is consistent with the phasing of vortex shedding above.

Similar observations as those made above concerning the effect of plus and minus waveforms at a reduced velocity of 5.16, also pertain in the case $U_\infty/f_eD = 5.89$ for which results are shown in Fig.10.

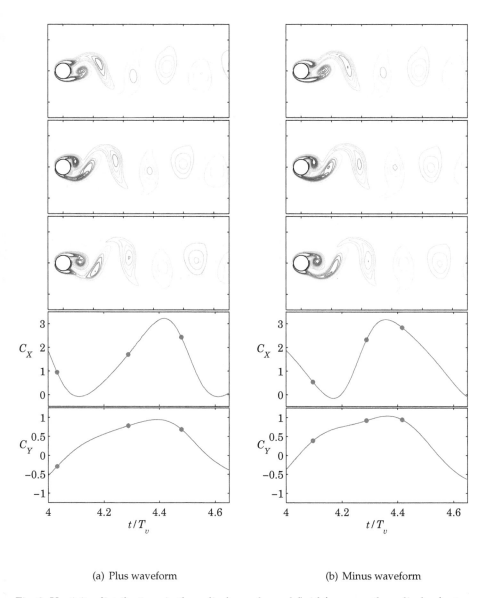

(a) Plus waveform   (b) Minus waveform

Fig. 9. Vorticity distributions in the cylinder wake and fluid forces on the cylinder for two different non-harmonic perturbation waveforms; $U_\infty / f_e D = 5.16$, $\Delta u / U_\infty = 0.20$, $Re \approx 150$.

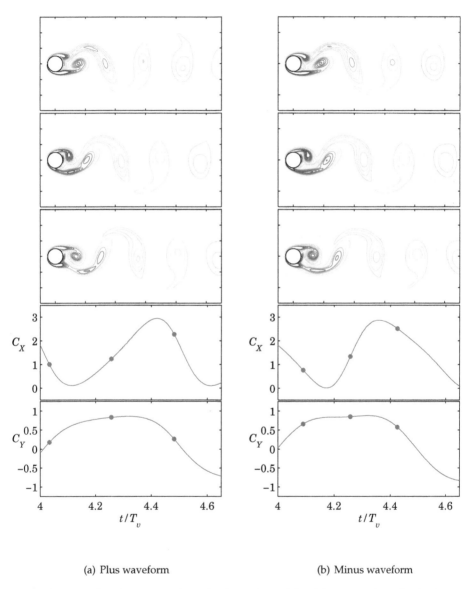

(a) Plus waveform   (b) Minus waveform

Fig. 10. Vorticity distributions in the cylinder wake and fluid forces on the cylinder for two different non-harmonic perturbation waveforms; $U_\infty / f_e D = 5.89$, $\Delta u / U_\infty = 0.20$, $Re \approx 150$.

	Plus waveform		Minus waveform	
$U_\infty / f_e D$	C_D	C_L'	C_D	C_L'
5.16	+3.9	+7.1	-3.9	-4.9
5.89	+2.7	+3.0	-2.0	-1.4

Table 2. Percent difference for two different non-harmonic waveforms compared to the fluid forces calculated for harmonically perturbed flow.

The differences in the time-averaged wake flow are very small for the two non-harmonic waveforms examined and need not be discussed in detail. On the other hand, the time-averaged mean and fluctuating forces reveal the global effect of the perturbation waveform. Table 2 shows the percent change compared to the harmonically perturbed flow for the same reduced velocity and amplitude ratio. The plus waveform causes an increase in the magnitude of both mean drag and rms lift and the opposite happens for the minus waveform. The change is not equivalent for the plus and minus waveform; it is larger for the plus waveform. In addition, these effects are more pronounced for $U_\infty / f_e D = 5.16$ (higher perturbation frequency) than for $U_\infty / f_e D = 5.89$.

6. Conclusion

In this chapter, issues related to the aerodynamics of bluff bodies in steady and time-dependent flows were presented. The subject was treated by numerical simulations of the 2D fluid flow around a circular cylinder which provided information on the dynamics of vortex formation and shedding in the wake and the induced forces acting on the body. Simulations were conducted for low Reynolds numbers in the laminar wake regime, mostly at $Re \approx 150$. Although this approach holds simplifying assumptions compared to real flows of practical interest, the observed phenomena convey over a wide range of Reynolds numbers and, in fact, makes it easier to discern the effects which are otherwise masked by wake turbulence and shear-layer instabilities that develop at higher Reynolds numbers. The results demonstrate that the vortex formation and the wake flow can be controlled by perturbations in the incident flow velocity. This control can be exercised by appropriately selecting the frequency and amplitude of the imposed perturbations on both the mean and fluctuating wake flow and, thereby, on the fluid forces on the cylinder. Whereas harmonic perturbations have been often utilized in the past, the present study also addresses the effect of non-harmonic perturbations, i.e. perturbations whose waveform deviates from a pure-tone harmonic. The findings of the present study illustrate that more effective wake control can be achieved by adjusting the waveform in addition to the perturbation frequency and amplitude.

7. References

Bearman, P. (1997). Near wake flows behind two- and three-dimensional bluff bodies, *Journal of Wind Engineering and Industrial Aerodynamics* Vol. 71: 33–54.

Gerrard, J. H. (1966). The three-dimensional structure of the wake of a circular cylinder, *Journal of Fluid Mechanics* Vol. 25(Part 1): 143–164.

Griffin, O. M. & Hall, M. S. (1991). Vortex shedding lock-on and flow control in bluff body wakes – Review, *Journal of Fluids Engineering* Vol. 113(No. 4): 526–537.

Henderson, R. (1995). Details of the drag curve near the onset of vortex shedding, *Physics of Fluids* Vol. 7(No. 9): 2102–2104.

Konstantinidis, E. & Balabani, S. (2007). Symmetric vortex shedding in the near wake of a circular cylinder due to streamwise perturbations, *Journal of Fluids and Structures* Vol. 23: 1047–1063.

Konstantinidis, E. & Balabani, S. (2008). Flow structure in the locked-on wake of a circular cylinder in pulsating flow: Effect of forcing amplitude, *International Journal of Heat and Fluid Flow* Vol. 29(No. 6): 1567–1576.

Konstantinidis, E. & Bouris, D. (2010). The effect of nonharmonic forcing on bluff-body aerodynamics at a low Reynolds number, *Jouranl of Wind Engineering and Industrial Aerodynamics* Vol. 98: 245–252.

Norberg, C. (2003). Fluctuating lift on a circular cylinder: review and new measurements, *Journal of Fluids and Structures* Vol. 17(No. 1): 57–96.

Papadakis, G. & Bergeles, G. (1995). A locally modified 2nd-order upwind scheme for convection terms discretization, *International Journal of Numerical Methods for Heat and Fluid Flow* Vol. 5(No. 1): 49–62.

Rhie, C. M. & Chow, W. L. (1983). Numerical study of the turbulent - Flow past an airfoil with trailing edge separation, *AIAA Journal* Vol. 21(No. 11): 1525–1532.

Triantafyllou, G., Kupfer, K. & Bers, A. (1987). Absolute instabilities and self-sustained oscillations in the wakes of circular-cylinders, *Physical Review Letters* Vol. 59(No. 17): 1914–1917.

Wu, M. H., Wen, C. Y., Yen, R. H., Weng, M. C. & Wang, A. B. (2004). Experimental and numerical study of the separation angle for flow around a circular cylinder at low Reynolds number, *Journal of Fluid Mechanics* Vol. 515: 233–260.

4

Simulation of Flow Control with Microjets for Subsonic Jet Noise Reduction

Maxime Huet, Gilles Rahier and François Vuillot
Onera – The French Aerospace Lab, F-92322 Châtillon
France

1. Introduction

With the increasing flight traffic in the past years grew up several types of pollutions for all near airports inhabitants. One of those major pollutions is the noise generated by aircrafts at takeoff. Despite all the work performed during the last decades on the understanding of the noise generation and its reduction, this nuisance still remains a really challenging problem. Above all the noise sources on a plane during takeoff, jet noise remains the dominant one, justifying all the efforts still performed on the topic.

For subsonic jets, radiated noise is a consequence of flow mixing. It is now universally admitted that this mixing noise is generated by the turbulence of the flow and that the noise producing region is axially restricted to about two potential core lengths (Fisher et al., 1977; Laufer et al., 1976). Turbulence is separated in this region in small turbulent eddies, with small dimensions compared to the nozzle, and large-scale structures. It is especially confirmed by the experimental observations (Tam et al., 2008) that both fine- and large-scale structures are the noise sources, large turbulent eddies being predominant in downstream direction especially for high speed jets.

Based on these observations, noise reduction devices are designed to act on jet turbulence development and especially to decrease the growth of the large scale structures in order to reduce major noise generation mechanisms. Several passive and active processes such as chevrons (Bridges & Brown, 2004; Nesbitt & Young, 2008) and microjets are currently being investigated experimentally and numerically. A major advantage of microjets, compared to passive devices, is their possibility to be turned off during cruise, which limits the thrust loss to takeoff configuration only, for instance.

The use of continuous air microjets on Mach 0.9 high Reynolds round jets has been experimentally investigated in the recent years (Alkislar et al., 2007; Arakeri et al., 2003; Castelain, 2006; Castelain et al., 2007; 2008). Measured data show that the interaction between each actuator and the jet shear layer corrugates the main flow over a distance of two to three jet diameters after the nozzle exit and generates a pair of counter-rotating axial vortices downstream of each microjet. All authors also observed a reduction of turbulence intensities in the shear layer and a lengthening of the potential core, with the exception of Alkislar et al. who noticed a turbulence increase 2 jet diameters after the nozzle exit, before its reduction downstream. In the far field, sound is decreased by 0.5 dB to 2 dB for all observation angles, which corresponds to low and medium frequency spectra reductions originating from the collapse of large turbulent structures. Higher frequency levels are increased by the microjets; this high frequency lift is explained by the enhancing of the small-scale structures by the fluid injection. Such turbulence modifications and related noise variations are similar to what is observed with chevrons, for instance (Alkislar et al., 2007).

A more important noise reduction up to 6 dB has been observed using water injection (Krothapalli et al., 2003), which did not exhibit high frequency lift. From these observations, Zaman (Zaman, 2010) suggests that this high frequency crossover is at least partly due to microjets self noise. Water injection nevertheless remains difficult to use on airplanes because of the important weight increase caused by the required water storage.

In addition to the above-mentioned actuators, fluidic control has also been investigated using pulsed microjets. When dealing with fluid injection, a major interest of such a control is the reduction of the injected mass flow rate. Ragaller (Ragaller et al., 2009) for instance demonstrated the capacity of achieving with a reduced mass flow pulsed control a noise decrease close to that obtained with continuous water injection.

Noise reduction through energy injection using plasma actuators is also being investigated for subsonic (Kastner et al., 2008; Kearney-Fischer et al., 2009a;b; Kim et al., 2009; Samimy et al., 2007a) and supersonic (Kearney-Fischer et al., 2011; Samimy et al., 2007b) jets. Control is made with localized arc filament plasma actuators, each actuator consisting in a pair of pin electrodes and generating electric discharge plasmas at a driven frequency varying from 0 to 200 kHz. Authors especially observed a broadband noise increase for low forcing frequency excitation, below $St_F = 1$, and a reduction for higher frequencies. This noise reduction is especially improved with increasing main jet temperature.

Only a limited number of numerical works have been published on the action of continuous air microjets for noise reduction (Enomoto et al., 2011; Laurendeau et al., 2008; Najafi-Yazdi et al., 2011; Rife & Page, 2011; Shur et al., 2011). Except for the simulation of Lew (Lew et al., 2010), they all modelled the microjets to avoid gridding the feed pipes. Those simulations illustrate the capacity of Large Eddy Simulations (LES) to capture the effect of microjets on both flow and noise, even using modelled actuators. To the knowledge of the authors, no simulations have been published on the action of pulsed actuators for jet noise control.

In the present work, a circular jet with an acoustic Mach number $M_a = U_j/c_0 = 0.9$ (U_j being the axial jet exhaust velocity and c_0 the ambient sound speed) is computed by LES. Simulations are performed for two main jet temperatures, corresponding to isothermal ($T_j = 288K$) and heated ($T_j = 576K$) flows with a Reynolds number based on nozzle diameter D of $Re_D = U_jD/\nu = 1,000,000$ and $320,000$ respectively, where ν stands for the kinematic viscosity.

The choice of those configurations is justified by the presence of aerodynamic and acoustic measurements previously performed without control by Institut PPRIME in Poitiers, France, during the JEAN European project, as well as aerodynamic and acoustic measurements realised by Castelain for the same nozzle and isothermal jet condition, with similar microjets configurations (Castelain et al., 2007; 2008). To complete those data, a new set of acoustic measurements has been conducted by Institut PPRIME during the project, to provide far field pressure data for the baseline and continuous microjets simulated configurations.

The paper is organised as follows. The test cases used for the simulations are recalled in section 2, with details given on the methodology used for both aerodynamic and acoustic simulations. Special focus is made on the modelling of the microjets. Section 3 corresponds to a detailed analysis of the aerodynamic fields. Mean and r.m.s. velocity profiles are presented and flow modifications with control are discussed. This part ends with a linear jet stability analysis. Acoustic results are then reproduced in section 4. Comparisons between simulations and experiments are made when available and changes in spectra shapes and integrated levels are analyzed. Relations are made with the evolutions previously noticed in the flow. Paper finally ends with concluding remarks and perspectives in section 5.

2. Simulation parameters

Aeroacoustics simulations are performed in two separate steps. Instantaneous flow simulations are conducted in a first time to compute the noise sources in the jet. To provide the best possible flow resolution with existing computed resources, accurate Navier-Stokes simulations are restricted to the jet plume, during which instantaneous aerodynamic fields are stored on surfaces surrounding the jet. Those fields are then used as source data for noise radiation to the microphones, using a surface integral method.

2.1 Aerodynamic numerical specifications

Aerodynamic simulations are performed using the flow solver CEDRE developed at Onera. CEDRE is a multi-physics, reactive solver used by researchers and aeronautical industries for engine conception and optimisation, such as combustion (Dorey et al., 2010; Dupoirieux & Bertier, 2011), turbine blade cooling (Guillou & Chedevergne, 2011) and jet noise (Bodard et al., 2009), for instance. Navier-Stokes equations are solved using second order upwind schemes space discretization for generalised polyhedral computational grids (Courbet et al., 2011) with explicit or implicit time schemes from first to third order for time resolution.

In the present simulations, the fluctuating jet flow is solved using the LES model with the MILES approach (Boris et al., 1992; Fureby & Grinstein, 1999; Grinstein & Fureby, 2002), where the dissipation of the structures smaller than the grid size is not modelled and is assumed to be of the order of the numerical dissipation. The validity of this approach for jet simulations relies on the hypothesis proposed by Biancherin (Biancherin, 2003) that those unresolved scales do not notably influence the noise generation in the flow. This assumption has been verified by Muller (Muller et al., 2006), who compared jet flow and noise results obtained using either the Smagorinsky subgrid scale model (Smagorinsky, 1963) or the MILES approach and observed very similar results.

The current approach developed for jet noise simulations is based on the inclusion of the nozzle geometry in the computational domain, which provides many advantages compared to simulations limited to flow domain only.

The first advantage is the absence of artificial disturbance at nozzle exit to destabilize the flow. Freund (Freund, 2001) and Bogey (Bogey et al., 2003), for instance, illustrated the necessity to add numerical perturbations over the imposed mean flow profile for the jet to be destabilized and to generate turbulence, with the risk of generating spurious noise. With the present approach, it is expected that by removing boundary conditions from a critical location, where instability waves must be allowed to develop freely, one permits some kind of natural growth of instability waves to occur. Indeed, Biancherin (Biancherin, 2003) and Muller (Muller, 2006) highlighted that, with this procedure, small truncation errors destabilize the flow and no more artificial excitation is needed at nozzle exit.

The inclusion of the nozzle all the more permits the simulation of complex, realistic geometries such as short-cowl nozzles with the inclusion of a pylon, bifurcations and chevrons (Eschricht

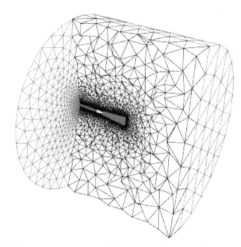

Fig. 1. Illustration of half the computational domain. Orange elements correspond to external boundaries, yellow to nozzle surfaces, blue to upstream fairing, red to structured patch and black to unstructured elements.

et al., 2008; Uzun & Hussaini, 2009; Xia et al., 2008; Yan et al., 2007). The increased difficulty in mesh generation is overtaken in the present case by the use of an unstructured grid generator.

The full numerical domain is a cylinder with length and radius of 120 D and 84 D respectively, mainly composed of tetrahedrons. Those important dimensions, far greater than the requested noise source generation domain, give possible the use of large sponge zones to avoid spurious noise reflections at the outlet boundaries of the computational domain, where static pressure is imposed. Inlet boundaries correspond to imposed total pressure and temperature at nozzle inlet and imposed velocity and static temperature at external flow inlet.

According to Zaman (Zaman, 1985), nozzle exit boundary-layer momentum thickness of a high Reynolds number jet of about 10^6 is $\delta_\theta/D \sim 10^{-3}$. Proper numerical resolution of such a thin boundary layer is unreachable with present computational power. The objective of the study being to illustrate the capacity of the simulations to quantitatively reproduce the noise reduction effect of microjets for industrial interests, it is chosen to restrict the grid size to affordable simulations with preferred high resolution of the flow in noise production regions. Boundary layers are thus under-resolved in the present computations, with the momentum thickness being roughly $\delta_\theta/D \sim 0.2$. This value is for instance four times larger than the one already used for other simulations of Mach 0.9 jets (Bodony & Lele, 2005; Bogey et al., 2003).

To increase resolution of the turbulence in the jet plume, a structured patch represented in red on Fig. 1 is included in the computational grid behind the nozzle exit. This truncated cone extends axially from $x/D = 0$ to $x/D = 25$, and radially to $r/D = 2$ and 4, respectively, at the two previous axial positions. Its construction is based on a 2D grid, rotated around the jet axis to provide 60 azimuthal hexahedral elements. The central part of the mesh is modified with a O-grid configuration illustrated on Fig. 2 to ensure homogeneous sized cells in the jet core.

The construction of the structured patch is based on aerodynamic and acoustic criteria to ensure sufficient resolution of flow development and sound propagation. First criterion ensures a sufficient resolution of the sheared flow and relies on the estimation of the jet

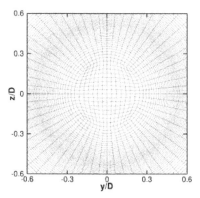

Fig. 2. Detail of the O-grid configuration used on jet axis. Axial position $x/D = 1$.

shear layer thickness δ by Candel and already used in previous simulations (Biancherin, 2003; Muller, 2006):

$$\frac{\delta}{D} = 0.153 \left(\frac{x}{D}\right) + 0.002 \tag{1}$$

Outside of the flow development region, an acoustic criterion is chosen to properly propagate acoustic waves to the storage surfaces used for noise radiation (see §2.3). For subsonic jets, where noise comes from flow mixing, dominant pressure waves are observed for Strouhal numbers $St = fD/U_j$ below 0.3. It is thus chosen to radiate noise with negligible dissipation up to $St = 0.5$, which provides most of the energy contained in the pressure spectra. Biancherin (Biancherin, 2003) demonstrated that, with the numerical methods used in the flow solver, a minimum of 20 points per wavelength is required for a correct propagation of the pressures waves over the distances considered with the constructed numerical grid.

In the present case, radial resolution is set to resolve the shear layer with 20 cells at an axial distance of 1 D after the nozzle exit, where the radial size of the cells is chosen to double between centreline and exterior of the shear with a regular growth of 7%. The same radial growth rate is used outside of the shear layer up to the end of the structured patch and satisfies the acoustic criterion for correct noise resolution. A 9.5% growth rate is used between the shear layer and the jet axis, where larger turbulent structures are expected to be observed especially after the end of the potential core.

Axial size of the elements is imposed at nozzle exit and is similar to the radial extent of the elements used in the shear layer at jet exhaust. Axial stretching of 1% is used until the elements reach the maximum size given by the acoustical criterion, after what no growth rate is applied.

The structured patch is composed of 1.6×10^6 hexahedra, for a total mesh of 4.5×10^6 cells. An illustration of the mesh at nozzle exit is given on Fig. 3 (a).

Construction of such grids leads to a limited amount of very small elements close to the nozzle exit, where velocity and temperature reach their maximum values. The use of an explicit time scheme would thus oblige one to run simulations with a very low time step verifying the CFL condition $(u + c)\,\Delta t/\Delta x < 1$, which can be very penalizing for efficient and quick simulations. Proper flow resolution in those elements not being critical for an accurate description of the whole jet, time resolution is performed with a first order implicit time scheme that allows CFL criterion above 1 in those cells, and thus an acceptable time step. Use of such a scheme is not a problem for noise propagation: results from previous

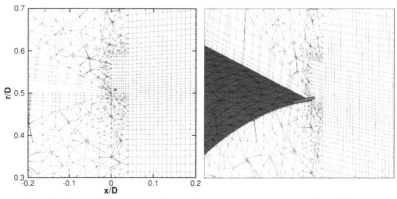

(a) detail of nozzle exit and mesh (b) 3D illustration of the cell used for
adaptation to model the microjets   microjet injection (in red)

Fig. 3. Mesh adaptation for microjets modelling. Planar cut of volume mesh is represented in
black, nozzle in grey and microjet cell in red.

studies demonstrated that, thanks to the very small time step in the acoustic propagation area
(chosen to verify a CFL condition lower than 1 in the flow), the first order implicit time scheme
is sufficient for a correct wave propagation and is consequently used in the present study.

2.2 Microjets modelling

To avoid the necessity of very small sized cells and restricting time step for the microjets
representation, a simplified model corresponding to volume sources located in specific cells is
employed. The objective of this approach is not to provide an extremely precise and detailed
description of the actuators. It is used to investigate the concept of flow injection for noise
reduction, which is numerically challenging and even innovative for pulsed actuators. One
must furthermore keep in mind that this approach should in the future be used for larger
scale, double stream configurations, for which a detailed description of the microjets will not
be numerically affordable in the near future due to the complexity of the configurations.

To reproduce the geometric configuration used experimentally, where the microjets are located
very close to the nozzle lip, the structured patch has been extended near the jet exit, as
illustrated on Fig. 3.

With the present approach, the mass injection and convection due to the microjets is modelled
through source terms in the equations of mass ($\dot{m}$), momentum ($\dot{m}\vec{U}$) and energy ($\dot{m}E$) in
specific hexahedral cells, identified as the locations of the microjets.

The choice is made here to calibrate the microjets with their velocity and temperature. Velocity
vector $\vec{U}$ corresponds to the orientation and velocity of the injected fluid and energy E is given
by the following formula:

$$E = h_{ref} + Cp\left(T - T_{ref}\right) + U^2/2 \qquad (2)$$

where T_{ref} is the reference temperature, h_{ref} the enthalpy of formation at reference
temperature and Cp the calorific capacity at constant pressure.

To ensure that the computed velocity and temperature correspond to the desired values, the mass flow rate $\dot{m}$ is determined through an iterative process. Based on the aimed velocity and temperature, an estimation of the initial mass flow rate can be calculated using the fluid velocity, its density and the surface S used for the injection: $\dot{m} = \rho U S$, where density is approximated by considering that the microjet flow occurs at ambient pressure. The surface can be evaluated from cell geometry and velocity vector: $S = \vec{U} \cdot \vec{S}/U$ with $\vec{S} = \sum_i S_i \cdot \vec{n}_i$ the sum of downstream exhaust surfaces S_i of the hexahedral cell used for microjet injection and $\vec{n}_i$ their associated outside-oriented normal vectors.

Using this iterative approach, mass flow rate is updated at each iteration until velocity and temperature reach the target values. From the computed surface used for the injection, it is in addition possible to evaluate the equivalent microjet diameter, considering a circular microjet nozzle : $d_{eq} = 2\sqrt{S/\pi}$.

2.3 Noise radiation

Instantaneous flow fields provided by the flow simulation and stored on surfaces are used to radiate sound to the microphones. Several integral formulations such as Lighthill (Lighthill, 1952), Kirchhoff (Lyrintzis, 1994) and Ffowcs Williams & Hawkings (FW-H) (Ffowcs Williams & Hawkings, 1969) are available to reconstruct far field noise from volume or surface fluctuations. It is usually preferred to use surface fields to limit the amount of stored data and numerical operations required to perform the noise radiation, which is done in the present study. Lighthill volume integral formulation can nevertheless be fruitful if one wants to identify the noise sources present in the flow (in Lighthill's sense), as done by Perez (Perez et al., 2007).

Based on the extended investigations performed on the topic by Rahier (Rahier et al., 2004), noise radiation is performed in the present case using FW-H porous surface formulation. In their paper, the authors especially demonstrated that this formulation makes it possible to place the control surface rather close to the jet, in a non-uniform flow. The surface position requirement is that the non-uniformities of the flow on the control surface are only due to convected gradients (such as density gradients related to thermal mixing in the case of a hot jet) and are not high sources of noise production (such as instability expanding or vortex pairing). On the contrary, the Kirchhoff method (with density or pressure as input data) is a much less suited tool for acoustic post-processing of jet aerodynamic simulations because of its high sensibility to non acoustic density or pressure gradients (local thermal mixing or vorticity).

Control surfaces are moreover kept open at both extremities. Investigations indeed showed that closing the surface leads in the best case to results similar as with open surface and can provide erroneous results if the control surface is too short, because of the turbulence level on the downstream closing disc. As a consequence of using an open control surface, its axial extent needs to be sufficiently large for a correct radiation at low angles, where the maximum noise is expected. It practically leads to a length of 20 to 25 nozzle diameters.

An axial extent of 25 diameters is used for the present simulations. Surfaces additionally verify $1 \leq r/D \leq 2$ at nozzle exit and $3 \leq r/D \leq 4$ at the downstream end. Radiation is performed for 4 different surfaces. Collapsing time signals and pressure spectra in the far field ensure that the surfaces enclose all noise sources and provide a limited dissipation for frequencies below a Strouhal number of 0.5.

T_s (K)	U_j (m/s)	M_a	M_j	Re_D	Q_j (kg/s)
288	306.65	0.90	0.90	1.0×10^6	0.723
576	306.65	0.90	0.64	3.2×10^5	0.362

Table 1. Main jets experimental characteristics.

To end, computed spectra and integrated pressure levels are azimuthally averaged over 48 microphones located around the jet axis. In the present case of an axisymmetric geometry, far field noise is indeed expected to be statistically independent of the azimuth, which is not the case for short-time duration computed signals and can lead to differences of almost 2 dB (Huet et al., 2009).

2.4 Simulated configurations and numerical procedure

The nozzle used for the simulations is a convergent single stream nozzle with an exhaust diameter $D = 50$ mm. Two main jet configurations, corresponding to isothermal and heated flow, are investigated, which main characteristics are recalled on Table 1. Those configurations correspond to the baseline and are used as reference when investigating the effect of microjets on flow and noise.

Based on grid construction and flow characteristics, the aerodynamic time step is set to $\Delta t_{aero} = 10^{-6}$ s, which corresponds to a CFL criterion lower than 1 in most of the computational domain. Surface storage is performed every 10 iterations of the flow simulation, $\Delta t_{acou} = 10^{-5}$ s. It corresponds to a maximum frequency of 50 kHz, which ensures that every frequency correctly resolved by the LES will be properly radiated using the integral formulation.

The control of the jet is performed with the use of 12 microjets regularly positioned around the nozzle lip, just after the exhaust area, with an impinging angle of 45 degrees relative to the jet axis. Continuous microjets are computed for both main jets. Pulsed microjets, for which simulations are run as exploratory computations only, are just performed for the isothermal configuration with all actuators in phase, which corresponds to the axisymmetric mode $m = 0$. Their flow and geometrical characteristics are identical to those used for the continuous microjets; time modulation of the source terms is a periodized crenel, whose value is set to 1 during $1/3^{rd}$ of the period and 0 otherwise. This modulation is a first representation of the plasma synthetic jets developed at Onera for flow and noise control (Caruana et al., 2009; Hardy et al., 2010). The two forcing frequencies $f_F = 3$ kHz and 9 kHz are considered for the simulations. They are chosen following the experimental results of Samimy (Samimy et al., 2007a) using localized arc filament plasma actuators and who observed a broadband noise increase for the low frequency excitation and a reduction for the high one.

For all simulations, microjets velocity and temperature are set to 300 m/s and 288 K respectively. Those values correspond to the ones used for the main isothermal jet.

Equivalent microjet diameter is $d_{eq} = 1.27$ mm. This diameter is comparable to the 1 mm diameter of the actuators used at Institut PPRIME during the acoustic test campaign of the project and by Castelain (Castelain et al., 2007; 2008) for aerodynamic and acoustic measurements performed in a separate framework. The distance between the microjets and the main jet shear layer is 0.40 d_{eq}. It ensures that the microjets flow spreading is very limited before reaching the jet flow and that its action on the shear layer is not negligible.

	x_c/D	$u'_{x\,rms}/U_j$ max at $r/D = 0$	$u'_{x\,rms}/U_j$ max at $r/D = 0.5$ ($x/D > 1$)
isothermal jet, baseline experiment	7.9	0.135	
heated jet, baseline experiment	6.3	0.155	
isothermal jet, baseline	5.2	0.167	0.186
isothermal jet, continuous microjets	5.7	0.159	0.183
isothermal jet, $f_F = 3$ kHz pulsed microjets	7.1	0.178	0.211
isothermal jet, $f_F = 9$ kHz pulsed microjets	6.6	0.153	0.176
heated jet, baseline	4.1	0.168	0.195
heated jet, continuous microjets	4.7	0.187	0.198

Table 2. Axial position of the end of the potential core x_c and peak axial r.m.s. velocity along the jet axis and the shear shear.

The computed injected mass for each continuous microjet is $Q_{mjet} = 4.8 \times 10^{-4}$ kg/s. This value is 5% more important than the estimation made at the beginning of the iterative process, based on fluid velocity, estimated density at ambient pressure and injection surface: $Q_{th} = 4.6 \times 10^{-4}$ kg/s. The small difference can come from pressure variations at microjets injection. The mass flow rate ratio per microjet $r_m = Q_{mjet}/Q_j$ is $r_m = 6.7 \times 10^{-4}$ for the cold jet and $r_m = 1.3 \times 10^{-3}$ for the heated one.

Microphones are located 50 diameters from the nozzle exit. JEAN experimental results, performed at a distance of 30 D, are rescaled to 50 D using the far field acoustic approximation with $1/r$ pressure amplitude decrease. Integrated pressure levels are computed for frequencies between 300 Hz and 41 kHz. The upper limit is higher than the grid cut-off frequency given in §2.1 because one of the objectives of the approach developed at Onera and used here is to compare simulations to experiments on the experimental frequency bandwidth. The underestimation of the computed OASPL caused by the poorly resolved high frequency levels is nevertheless negligible because the major part of the energy is contained in the low frequency range, below $St = 0.5$, that is accurately resolved in the simulations.

The numerical procedure used for all simulations is the following. Every aerodynamic simulation starts from rest and inlet pressure is progressively increased to reach the target value. Computation is then run until the flow is developed in the refined domain, after what surface storage and mean flow averaging begins. The storage is performed during 50 ms of simulated time, which corresponds to 300 D/U_j convective time units. This duration ensure statistically converged mean flow fields and sufficiently resolved far field pressure spectra in the frequency domain for further analyses.

3. Aerodynamic results

3.1 Reference simulations

Evolution of mean axial velocity and turbulent axial velocity along the jet axis are represented on Fig. 4 for the reference simulations. For both temperatures the potential core length, defined as the axial distance from the nozzle exit at which $U_x/U_j = 0.9$ and reported on Table 2 is about 30% shorter in the simulation compared to the experiment. Both simulations moreover overestimate the peak axial r.m.s. velocity on the jet axis.

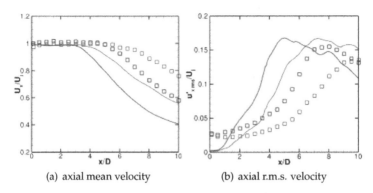

(a) axial mean velocity   (b) axial r.m.s. velocity

Fig. 4. Experimental and simulated axial (a) mean and (b) r.m.s. velocity on jet axis for isothermal (blue) and heated (red) baseline configurations. □ experiments; – simulations.

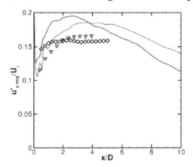

Fig. 5. Evolution of the axial r.m.s. velocity u'_{rms} along the nozzle lip. Simulations: – isothermal and – heated baseline configurations. Measurements: ◊ Husain and Hussain for an initially turbulent shear layer (Husain & Hussain, 1979) and ∇ Fleury et al. for a Mach 0.9 jet at $Re_D = 7.7 \times 10^5$ (Fleury et al., 2008).

This numerical underestimation of the potential core length has already been observed in previous computations of the same geometry (Andersson et al., 2005; Bodard et al., 2009; Bogey & Bailly, 2006) and might come partly, in the present computations, from different initial conditions between experiments and simulations. Fig. 4 (b) illustrates the differences between experimental and simulated turbulence levels at nozzle exit, experiments exhibiting the presence of velocity perturbations which are not present in the computations. These very low computed turbulence levels are a consequence of the unperturbed inflow imposed on the nozzle inlet boundary and may be the reason of the discrepancies observed. Bogey (Bogey & Bailly, 2010) indeed numerically demonstrated for a $M_j = 0.9$ and $Re_D = 10^5$ jet that the presence of disturbances in the nozzle increase the potential core length and lower the turbulence levels on the jet axis.

Simulated turbulence profiles along the nozzle lip reproduced on Fig. 5 exhibit a peak value for $x/D \sim 4$ and $x/D \sim 3$ respectively for the isothermal and the heated jet, whereas those turbulence levels increase nearly monotonically for experimental fully turbulent jets also reported on the figure. These peak r.m.s. values might indicate a strong transition of the initially laminar jet mixing layer in the computations.

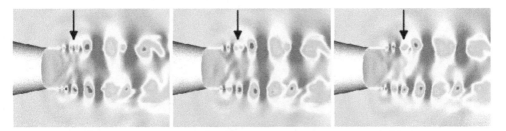

Fig. 6. Illustration of the vortex pairing phenomenon (arrow) observed for the isothermal simulated jet. Coloured map represents density field at 3 different instants.

With the considered Reynolds numbers, the shear layer is expected to be fully turbulent for the isothermal jet or at least nominally turbulent for the heated jet, using the terminology proposed by Zaman (Zaman, 1985). Its initially laminar state might be explained by the unperturbed imposed inflow as well as the lack of resolution of the boundary layers of the nozzle. This low discretization would not allow sufficient resolution of the perturbations that are expected to develop and to trigger the turbulent shear layer development. Similar observations have been made by Vuillot (Vuillot et al., 2011) on a double stream nozzle with external flow.

An illustration of the initially laminar shear layer development is visible on Fig. 6 for the isothermal jet. One can first observe on the instantaneous density field represented at three different instants a vortex shedding phenomenon, corresponding to the green dots along the shear layer. This shedding is followed by multiple vortex pairings occurring 1 diameter or more after the nozzle exit, one of them being visible in the pictures at the end of the arrow. These phenomena are characteristic of laminar to turbulent transition in the mixing layer, explaining the peak axial r.m.s. velocity observed on Fig. 5 for the computations.

In addition to the initially laminar state of the computed jets, the overestimation of the turbulence observed on Fig. 4 (b) and on Fig. 5 might also come partly from a too fast coarsening of the grid in the axial direction. Indeed, a coarse grid tends to overestimate the large turbulent structures and thus the turbulent kinetic energy, leading to a too fast growth of the shear layers and a shortening of the potential core.

Despite those discrepancies between the simulations and the experiments, the effect of temperature on mean and r.m.s. velocities is well reproduced by the simulations, with a reduction of the potential core length of 20% with increasing temperature for both measurements and computations. The consequence of this core length shortening is the reduction of the axial distance at which the peak axial r.m.s. velocity occurs. The peak axial r.m.s. level increase observed experimentally on the jet axis with increasing temperature is nevertheless not clearly visible in the simulations, with only a 0.5% raise compared to 20% in the experiments. Higher turbulence levels for the heated jet are however observed numerically along the shear layer.

3.2 Continuous microjets

No aerodynamic measurements being available for the simulated configurations with microjets, investigations presented hereafter are performed comparing baseline and controlled simulation with available literature results.

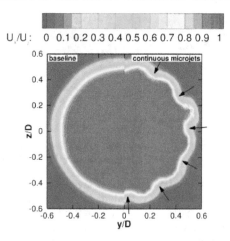

Fig. 7. Mean axial velocity computed at $x/D = 1$ for the baseline (left) and continuous microjets (right) isothermal configurations. Arrows represent the position of the microjets.

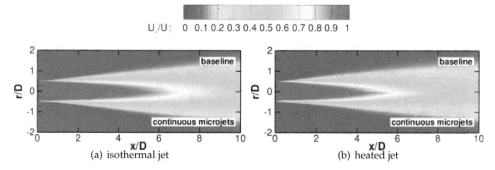

Fig. 8. Mean axial velocity distribution for (top) baseline and (bottom) continuous microjets configurations. Plane located through microjet position.

An illustration of the efficiency of the microjets for modifying main jet mean flow is visible on Fig. 7. It is clearly visible on this figure that the actuators corrugate the jet azimuthally by locally decreasing the axial velocity downstream of the microjets. These corrugations are limited in space to the nozzle vicinity and vanish more downstream after 3 diameters, as experimentally observed (Alkislar et al., 2007; Arakeri et al., 2003; Castelain et al., 2007). They are a consequence of the difference in axial velocity between the microjets and the main jet that is expected to generate flow mixing.

Cartographies of mean axial velocity, reproduced on Fig. 8, illustrate a stretching of the jet with the presence of the microjets. This stretching is accompanied by an increase of the potential core length of 10% and 15%, respectively for the isothermal and heated configurations, especially visible on Fig. 12 (a). This result is consistent with the experimental observations of Castelain et al. and Arakeri et al., the latter observing an increase of 25% caused by the control.

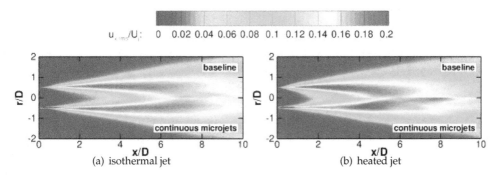

Fig. 9. Axial r.m.s. velocity distribution for (top) baseline and (bottom) continuous microjets configurations. Plane located through microjet position.

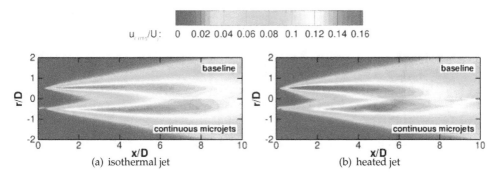

Fig. 10. Radial r.m.s. velocity distribution for (top) baseline and (bottom) continuous microjets configurations. Plane located through microjet position.

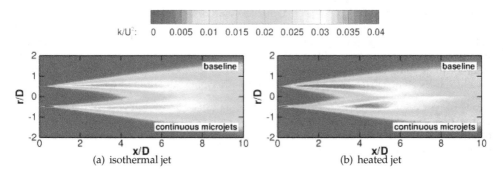

Fig. 11. Turbulent kinetic energy distribution for (top) baseline and (bottom) continuous microjets configurations. Plane located through microjet position.

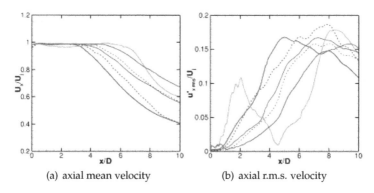

(a) axial mean velocity   (b) axial r.m.s. velocity

Fig. 12. Simulated axial (a) mean and (b) r.m.s. velocity on jet axis. Isothermal jets: – baseline; – – continuous microjets; – $f_F = 3$ kHz pulsed microjets; – $f_F = 9$ kHz pulsed microjets. Heated jets: – baseline; – – continuous microjets.

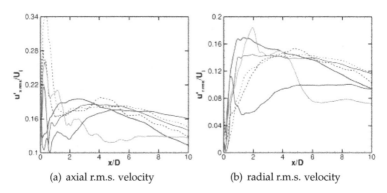

(a) axial r.m.s. velocity   (b) radial r.m.s. velocity

Fig. 13. Evolution of the r.m.s. velocities along the nozzle lip. The legend is the same as on Fig. 12.

As experimentally expected, the presence of the microjets also leads to a reduction of the peak axial r.m.s. velocity on the jet axis for the isothermal jet, see Fig. 12 (b). This peak value is however increased for the heated jet and is a consequence of a spot of high axial velocity fluctuations appearing near the jet axis for $x/D = 8$, as visible on Fig. 9. Except for this spot, no significant modification of the peak value is observed for the heated jet, its axial location being shifted downstream compared to the baseline configuration because of the more important jet stretching relative to the isothermal simulations.

Along the shear layer and for the isothermal jet, actuators lead to a reduction of the turbulence in the first diameters after the nozzle exit, see Fig. 13. This result is similar with the observations of Castelain (Castelain, 2006). It nevertheless slightly differs from the results of Arakeri (Arakeri et al., 2003) who measured a reduction of the peak axial r.m.s. velocity that is not reproduced by the simulations. Similar flow modifications are observed for the heated jet, except for the peak radial r.m.s. velocity that is now reduced in the computations.

Such modifications in the turbulent velocities numerically lead, for both temperatures, to a reduction of the peak turbulent kinetic energy in the jet plume, that appears further

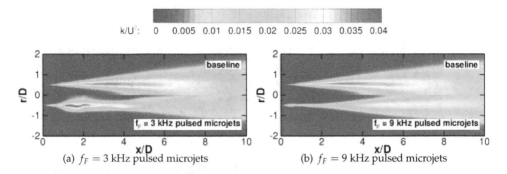

Fig. 14. Turbulent kinetic energy distribution for (top) baseline and (bottom) pulsed microjets configurations. Plane located through microjet position.

downstream from the nozzle exit compared to the baseline configurations, as visible on Fig. 11. On this figure is also observed a high turbulence level just downstream of the microjets injection, that comes essentially from the contribution of the axial turbulent velocity. This increase is not observed in the experiments and can be attributed in the present simulations to the initially laminar shear layer.

3.3 Pulsed microjets

Despite the initial state of the shear layer that is not fully turbulent in the simulations, which leads especially to some differences on flow characteristics with the experiments, all the aerodynamic simulated results with continuous microjets presented in the section above compare rather well with the experimental data. This tends to illustrate that the actions of the actuators on the main jet are correctly reproduced in the computations. A further investigation on the efficiency of such a type of control is thus performed by considering pulsed microjets. It has to be recalled nevertheless that the modelled pulsed actuators do not intend to precisely reproduce the existing plasma synthetic jets developed at Onera (Caruana et al., 2009; Hardy et al., 2010), but are used to illustrate the capacity of the numerical code to reproduce the action of such microjets.

Fig. 12 illustrates the more important potential core length increase obtained with the pulsed control, compared to the continuous one. This core increase is related for both simulations to the reduction of the axial r.m.s. velocity along the shear layer, the more important reduction being observed for the high frequency excitation. This turbulence reduction observed for the $f_F = 9$ kHz forcing is qualitatively similar to the one occurring with continuous microjets, for both axial and radial turbulent velocities, and is simply more pronounced, as visible on Fig. 13. This is clearly visible when looking at the turbulent kinetic energy distribution, reproduced on Fig. 14 (b).

A different evolution of the turbulence is observed for the low frequency forced jet. Velocity fluctuations present in this case a very large increase a few diameters after the nozzle exit, the peak turbulent kinetic energy being localized about 1.5 D downstream of the jet exit, see Fig. 14 (a). The oscillations observed on this figure only come from the too short time duration in the averaging and are caused by the strong response of the jet to the excitation, visible on Fig. 15 (c), and which was not observed with continuous or $f_F = 9$ kHz pulsed control.

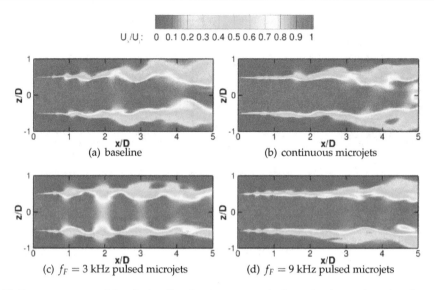

U_x/U_j: 0 0.1 0.2 0.3 0.4 0.5 0.6 0.7 0.8 0.9 1

(a) baseline

(b) continuous microjets

(c) $f_F = 3$ kHz pulsed microjets

(d) $f_F = 9$ kHz pulsed microjets

Fig. 15. Instantaneous axial velocity distribution on $y = 0$ plane. Isothermal jet simulations.

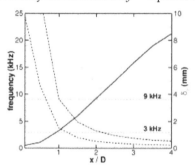

Fig. 16. Evolution of the (–) momentum thickness, (– –) most unstable shear layer frequency and (– ·) neutral disturbance frequency. Baseline isothermal jet simulation.

To explain this distinct response of the flow to the $f_F = 3$ kHz pulsed excitation, a jet stability analysis is performed on the baseline configuration, using Michalke's theoretical work on the stability of inviscid shear layers (Michalke, 1965). In this theory, the radial profile of the axial velocity is analytically represented with an hyperbolic-tangent profile:

$$U(r) = \frac{U_j}{2}\left(1 + tanh\left(\frac{r}{2\delta_\theta}\right)\right) \tag{3}$$

where r represents the radial distance from the axis and δ_θ the momentum thickness. The most unstable shear layer frequency is then given by $f = 0.0165 \cdot U_j/\delta_\theta$ and the neutral disturbance frequency by $f = 0.0400 \cdot U_j/\delta_\theta$. This last frequency corresponds to the threshold value for which the excited shear layer modes are stable and thus do not lead to a destabilization of the jet.

In the present simulation, the hyperbolic-tangent velocity profiles well fit the simulated profiles in the first jet diameters. More downstream some differences can be observed between simulated and analytical profiles, that might lead to an underestimation of the calculated frequencies. The modelling however gives a correct order of those two characteristic frequencies of the shear layer.

The evolution of the momentum thickness in the first diameters after the nozzle exit is represented on Fig. 16. On this figure are also visible the most unstable shear layer frequency and the neutral disturbance frequency. It is clearly observed that both frequencies of 3 kHz and 9 kHz correspond to most unstable frequencies within the first diameter after the nozzle exit. However, the neutral frequency is observed just downstream for the 9 kHz frequency, which can explain that no destabilization is noticed: each instability appearing in the jet is nearly instantaneously damped. This damping is theoretically expected further in the jet for 3 kHz perturbations, which gives the time for the shear layer instabilities to develop. This result is coherent with the illustration of the instantaneous axial velocity on Fig. 15 (c), where strong oscillations are observed with the $f_F = 3$ kHz pulsed control for $x/D = 1$.

4. Acoustic results

Noise modifications caused by temperature increase and microjets are presented in this section. Simulations with continuous control are compared to measurements provided in the frame of the study by French institute PPRIME for angles above 50 degrees. Experimental data from European project JEAN, also performed by institute PPRIME, are nevertheless preferred for temperature effect investigation, as they provide noise levels below 50 degrees. Comparisons have been made to ensure that both experimental campaigns provide similar results.

4.1 Temperature effect

The effect of temperature on jet noise has been widely investigated experimentally (Tanna, 1977; Tanna et al., 1975; Viswanathan, 2004). It has been observed that, for a fixed acoustic Mach number $M_a = U_j/c_0$, where c_0 is the sound velocity at ambient temperature, the radiated noise directly depends on the jet temperature. For $M_a < 0.7$, increasing the jet temperature increases the noise; no effect is observed for $M_a = 0.7$ and the noise is reduced with increasing temperature for $M_a > 0.7$. For this last case corresponding to the configurations presented here, the spectral modifications are a decrease of the medium and high frequency levels when temperature rises, while no modifications are visible for the low frequency part of the spectra.

Experimental data reproduced on Fig. 17 exhibit spectral modifications with temperature increase that conform to literature observations. One can nevertheless notice a slight noise increase around 1 kHz at 60 degrees, which was not mentioned in the above-cited published results.

Simulated power spectral densities are illustrated on Fig. 18. Except for very low frequencies, where jet heating reduces the noise levels, calculations reproduce rather well the effect of temperature experimentally described. The decrease of medium and high frequency levels with increasing temperature is especially well observed, as well as the small level increase for the heated jet near 1 kHz at 60 degrees.

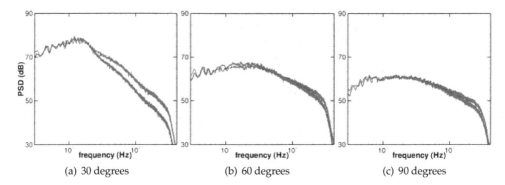

(a) 30 degrees   (b) 60 degrees   (c) 90 degrees

Fig. 17. Illustration of the temperature effect on power spectral densities for the experimental baseline configurations. – isothermal jet; – heated jet.

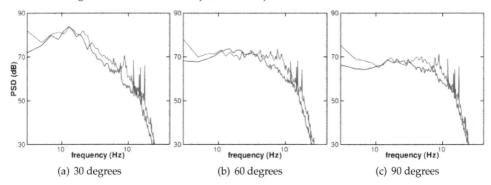

(a) 30 degrees   (b) 60 degrees   (c) 90 degrees

Fig. 18. Illustration of the temperature effect on power spectral densities for the simulated baseline configurations. – isothermal jet; – heated jet.

Comparing each simulation with the corresponding experiment, see Figs. 19 & 20, one first observes tonal noises in the computations for frequencies above 10 kHz. Those peaks are not observed experimentally and come from the vortex shedding and multiple pairings previously evidenced in §3.1. They are thus a consequence and an illustration of the initially laminar development of the simulated jets. It is coherent with the linear stability analysis presented above; the frequencies of these tonal noises are indeed close to the most unstable shear layer frequency in the vicinity of the jet exhaust, as visible on Fig. 16. These tones are nevertheless not very energetic and do not significantly contribute to the integrated levels.

Simulated PSD also overestimate experiments by about 5 dB, with for instance a maximum level of 84 dB at the peak frequency in both simulations, compared to 79 dB for the measured data. These higher calculated levels can be explained by the overestimation of the turbulent kinetic energy in the jet, as discussed in §3.1. Numerical spectra nevertheless qualitatively well collapse with experiments especially at low angles, for which the frequency of maximum level is particularly very well reproduced. The important levels observed numerically around 5 kHz at 90 degrees, particularly visible for the isothermal jet, are explained by the initially laminar jet, which generates an additional sound source compared to an initially turbulent jet (Bogey & Bailly, 2010).

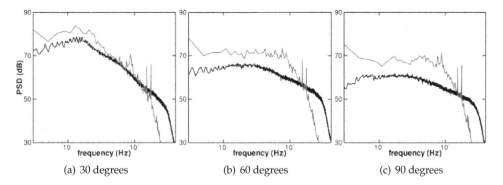

(a) 30 degrees   (b) 60 degrees   (c) 90 degrees

Fig. 19. Comparison of experimental and computed power spectral densities for the isothermal jet. – experiments; – simulations.

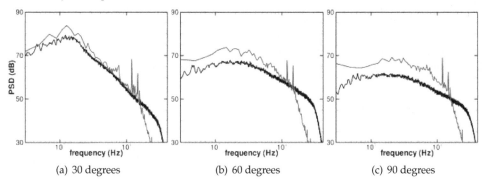

(a) 30 degrees   (b) 60 degrees   (c) 90 degrees

Fig. 20. Comparison of experimental and computed power spectral densities for the heated jet. – experiments; – simulations.

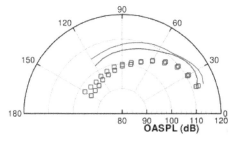

Fig. 21. Experimental and computed integrated levels for isothermal (blue) and heated (red) baseline configurations. □ experiments ; – simulations.

As a consequence of the observations made on PSD, simulated OASPL, represented on Fig. 21, overestimate the measurements at all angles of about 6 dB. Directivity patterns are nevertheless well recovered in the computations, with for instance the maximum level found at about 30 degrees for all configurations. The medium and high frequency levels reduction observed in the spectra for the heated jet lead to lower integrated noise levels for almost all angles. Near 50 degrees, however, experimental OASPLs are higher for the hot jet and are very close between the two temperatures for the simulations. This difference is caused by the more important noise increase near 1 kHz observed experimentally. To end, the reduction is more pronounced in the simulations with a 2.3 dB difference at 90 degrees, for example, compared to 0.7 dB in the experiments.

The simulated acoustic results obtained for both jets are coherent with the aerodynamic observations made previously. A good qualitative agreement is made with the experiments, the overestimation of the radiated sound being essentially caused by the higher simulated turbulence levels and the initially laminar jet development. The good agreement between experiments and simulations finally illustrates that jet physics and especially the sound sources are correctly reproduced in the simulations for both temperatures and allow the investigation of noise reduction with microjets.

4.2 Continuous microjets

Experimental power spectral densities without and with continuous microjets are reproduced on Figs. 22 & 23 for the isothermal and heated jets, respectively, and integrated levels are visible on Fig. 24. Spectra present high frequency bumps at 60 degrees, caused by reflection on the microphone that is located too close to the acoustically treated ground. Those bumps however occur for levels far below the peak ones and thus do not contaminate the integrated pressure levels. They are moreover high frequency perturbations above the accurately resolved numerical frequencies and do not prevent the relative comparisons of the spectra without and with microjets for energetic levels.

For both temperatures, the noise modifications caused by the microjets are similar to the literature results of Alkislar (Alkislar et al., 2007) and Castelain (Castelain et al., 2008) for isothermal jets, with essentially the most important spectral reduction observed close to the peak frequency and a reduction of the OASPL levels over the entire range of measurement angles caused by the medium frequency noise decrease. The integrated noise reduction of 1.1 dB observed at 90 degrees is besides very close to that of 1.2 dB measured by Castelain et al. on a similar configuration and with the same mass flow rate ratio per microjet $r_m = 6.7 \times 10^{-4}$.

For the two temperatures, simulated spectral modifications caused by the control are similar to the ones observed in the experiments, see Figs 25 & 26. With the exception of the heated jet at fore angles, where low frequency noise is increased by the presence of the microjets, the low frequency parts of the spectra are barely modified while medium frequency levels are reduced by the actuators. The control nevertheless leads to a higher pressure levels reduction in the simulations, with a noise decrease of more than 2 dB at 60 degrees, for instance, compared to a maximum of 1 dB in the measurements. The minimum frequency for which the noise reduction is observed is also higher in the simulations, with numerically a noticeable attenuation for frequencies above several kHz only, while such a reduction is already observed below 1 kHz experimentally.

This reduction of the pressure levels can be related to the decrease of the turbulence fluctuations in the jet detailed in §3.2. The discrepancies observed can be explained by jet initial state at nozzle exit. It can indeed be expected that the control, designed to favour the collapsing of large turbulent structures, is more effective with the coherent structures present in the initially laminar shear layer than in the turbulent shear layer. Noise reduction might thus be overestimated in the simulations compared to the experiments.

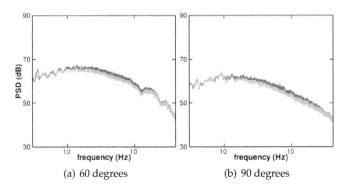

(a) 60 degrees   (b) 90 degrees

Fig. 22. Experimental power spectral densities. – baseline; continuous microjets. Isothermal main jet configuration.

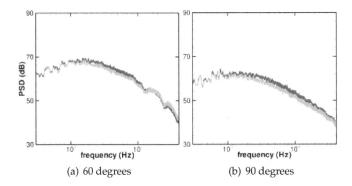

(a) 60 degrees   (b) 90 degrees

Fig. 23. Experimental power spectral densities. – baseline; continuous microjets. Heated main jet configuration.

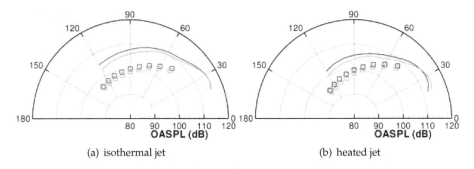

(a) isothermal jet   (b) heated jet

Fig. 24. Experimental and computed integrated levels for experimental and simulated (a) isothermal and (b) heated configurations without and with continuous microjets.
□ experiments; – simulations. Isothermal baseline in blue, heated baseline in red, continuous microjets in green.

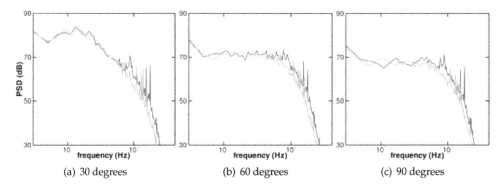

(a) 30 degrees   (b) 60 degrees   (c) 90 degrees

Fig. 25. Simulated power spectral densities. – baseline; continuous microjets. Isothermal main jet configuration.

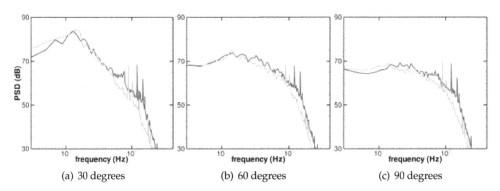

(a) 30 degrees   (b) 60 degrees   (c) 90 degrees

Fig. 26. Simulated power spectral densities. – baseline; continuous microjets. Heated main jet configuration.

To end with the spectra modifications, the noise increase at high frequency is not reproduced numerically, because those frequencies and the corresponding noise sources are above the grid cut-off frequency and are thus not properly resolved in the simulations. Not taking into account those high frequency levels is nevertheless not penalizing when computing the integrated pressure levels because those levels are much lower than the most energetic ones and thus have a negligible contribution to the OASPL, as said previously in §2.4.

Integrated pressure levels of the isothermal simulations, visible on Fig. 24 (a), present a noise reduction of 1.1 dB and 1.8 dB, respectively at 60 and 90 degrees. The latter reduction is more important than the experimental one and is a consequence of the larger medium frequency levels reduction with microjets in the simulations, discussed above. Similar OASPL reductions are observed above 50 degrees for the heated jet, with a numerical noise decrease of 1.5 dB at 90 degrees, with comparison to 1.4 dB in the experiments. In this last case, the larger medium frequency noise reduction observed numerically is partly compensated by the low frequency noise increase caused by the control, that finally leads to very close simulated and measured integrated levels.

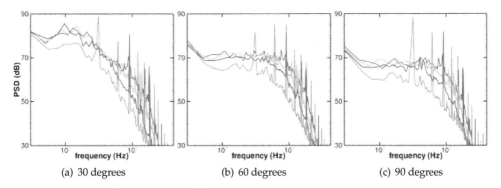

(a) 30 degrees   (b) 60 degrees   (c) 90 degrees

Fig. 27. Simulated power spectral densities. – baseline; continuous microjets; $- - f_F = 3$ kHz pulsed microjets; $- - f_F = 9$ kHz pulsed microjets. Isothermal main jet configuration.

No experiments are available close to the jet axis but results from the literature (Alkislar et al., 2007; Castelain et al., 2008) indicate, for isothermal jets, that noise reduction is also expected at fore angles with for instance an attenuation of 0.8 dB at 30 degrees for a configuration similar to the isothermal jet considered in the present study. Such a reduction is not reproduced numerically even if noise levels are reduced above 7 kHz and 4 kHz, respectively for the isothermal and the heated jets. For such angles low frequency noise, around 1 kHz, strongly dominates the spectra and is not reduced by the presence of the microjets. Thus, the medium frequency levels reduction provided by the control is not significant on the integrated levels. This low frequency noise is even increased for the heated jet and leads to a higher OASPL at 20 degrees with microjets, compared to the baseline configuration. A possible reason for experimental and numerical noise discrepancies, jet laminar or turbulent initial state, has already been discussed previously.

4.3 Pulsed microjets

The spectral densities obtained with pulsed control are reproduced on Fig. 27 where are also represented the baseline and continuous microjets configurations spectra. The two principal modifications caused by the pulsed microjets are first the presence of tonal noises at the harmonics of the forcing frequency and second the higher broadband noise reduction compared to continuous control.

The large broadband noise reduction is coherent with the aerodynamic observations and can be linked to the important decrease of the turbulent velocities along the shear layer with the periodic control. Despite the fact that jets developments have been found to be very different with the two forcing frequencies, the broadband noise reduction is similar for the two simulations and corresponds to a large level decrease for almost all frequencies, except for the very low part of the spectra that are not modified by the control. The highest reduction is observed for the $f_F = 3$ kHz forcing with a decrease over 10 dB for a wide frequency range, for which the largest axial turbulent velocity reduction was observed in the shear layer.

The presence of tonal noises is coherent with available literature results and cannot be attributed only to the rough model used for the microjets. Samimy (Samimy et al., 2007a) for instance experimentally observed similar spectral peaks using localized arc filament plasma actuators. Broadband noise modification with the two different forcing frequencies

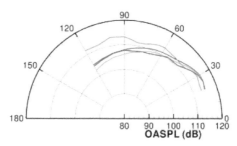

Fig. 28. Simulated integrated pressure levels. – baseline; continuous microjets; – – $f_F = 3$ kHz pulsed microjets; – – $f_F = 9$ kHz pulsed microjets. Isothermal main jet configuration.

is nevertheless different to that observed by Samimy where a broadband noise increase was observed for the low forcing frequency $f_F = 3$ kHz and a reduction noticed for the high forcing frequency $f_F = 9$ kHz. This difference might come from the different types of actuators used in their measurements and in the present simulations. Experimentally, the control is made through the injection of pure energy while the microjets used numerically correspond to the injection of mass and its convection. It can also be a consequence of the initially laminar state of the simulated shear layer.

If confirmed by further simulations and experiments, those results are of important interest in the use of pulsed control for jet noise reduction. Nevertheless, the gain obtained in the present simulations thanks to the broadband noise reduction is annihilated by the strong contribution of the tonal noises, which finally leads to integrated levels of the same order or greater than that of the baseline configuration for almost all angles, as illustrated on Fig. 28. One must keep in mind that pulsed microjets simulations performed here are a first attempt to illustrate the capacity of the numerical codes to reproduce the actions of such actuators on flow development and noise generation. Flow and noise investigations need to be continued and microjets modelling to be improved to increase the fidelity of the simulations, to evaluate a possible suppression or reduction of the tonal noises or to identify the numerical contribution of the microjets modelling in the far field noise radiation, for instance.

5. Conclusion

Simulations of an isothermal and a heated jet with an acoustic Mach number of 0.9 are presented. Flow and noise modifications caused by continuous microjets are investigated and compared to available experimental data. Noise reduction with pulsed actuators is also investigated for the isothermal configuration.

Simulated baseline configurations present a correct aerodynamic agreement with the experiments. An overestimation of the turbulence in the jet plume is nevertheless observed and can be explained essentially by the different initial state of the shear layer between experiments and simulations. This difference in the initial state, numerically laminar and experimentally turbulent might come from the numerical boundary condition in the nozzle where no flow perturbations are seeded. Far-field pressure spectra are in a good qualitative agreement with the experiments, except for the presence of tonal noises that are a consequence of the jet initially laminar state. Measured directivity patterns are also well recovered in the simulations, with however an overestimation of the absolute levels because of the excessive turbulence in the computed flow.

Flow and noise modifications with increasing temperature are also well reproduced numerically and indicate that the flow physics is well captured in the simulations. Similar conclusions are drawn with the presence of the continuous microjets that notably lead to an axial stretching of the flow fields and a reduction of the medium and high frequency noise of about 1.5 dB at 90 degrees. To end, a drastic turbulence reduction is observed numerically in the shear layer with the use of the pulsed microjets and is accompanied by an important broadband noise reduction in the far field. This noise decrease is nevertheless penalized by important tonal noises at the harmonics of the excitation frequency that makes this control unsuitable for noise reduction as is. Tonal noises have also been observed experimentally for a different type of pulsed actuators and measurements are now required with periodic fluid injection to confirm the promising broadband noise reduction of this type of periodic control.

From all the results presented above, the correct relative effect of the microjets observed numerically on the pressure spectra illustrate the capacity of the simulations to reproduce the action of the control on the flow development, despite the simplified representation of the microjets. Following works will have to focus at first on the simulation of initially turbulent simulated jets to improve flow and noise computation and to reduce discrepancies with the experiments. Improvement of microjets modelling will also be continued, for the pulsed actuators to be more representative of existing experimental devices and also to allow simulations of continuous and periodic controls on larger scale models. The final objective of these future studies is to identify the more promising noise reduction methodology and to help designing optimised experimental devices.

6. Acknowledgments

This work has been conducted within the French project OSCAR (Optimisation de Systèmes de Contrôle Actifs pour la Réduction du bruit de jet) managed by Onera and funded by the FRAE. The authors gratefully acknowledge P. Jordan from Institut PPRIME for providing the JEAN experimental results.

7. References

Alkislar, M. B., Krothapalli, A. & Butler, G. W. (2007). The effect of streamwise vortices on the aeroacoustics of a Mach 0.9 jet, *Journal of Fluid Mechanics* Vol. 578: pp. 139–169.

Andersson, N., Eriksson, L.-E. & Davidson, L. (2005). Large-Eddy Simulation of subsonic turbulent jets and their radiated sound, *AIAA Journal* Vol. 43(No. 9): pp. 1899–1912.

Arakeri, V. H., Krothapalli, A., Siddavaram, V., Alkislar, M. B. & Lourenco, L. M. (2003). On the use of microjets to suppress turbulence in a Mach 0.9 axisymmetric jet, *Journal of Fluid Mechanics* Vol. 490: pp. 75–98.

Biancherin, A. (2003). *Simulation aéroacoustique d'un jet chaud subsonique*, PhD thesis, Université Paris VI.

Bodard, G., Bailly, C. & Vuillot, F. (2009). Matched hybrid approaches to predict jet noise by using Large-Eddy Simulation, *Proceedings of the 15th AIAA/CEAS Aeroacoustics Conference*, number AIAA 2009-3316.

Bodony, D. J. & Lele, S. K. (2005). On using Large-Eddy Simulations for the prediction of noise from cold and heated turbulent jets, *Physics of Fluids* Vol. 17: 085103.

Bogey, C. & Bailly, C. (2006). Computation of a high Reynolds number jet and its radiated noise using Large Eddy Simulation based on explicit filtering, *Computers and Fluids* Vol. 35: pp. 1344–1358.

Bogey, C. & Bailly, C. (2010). Influence of nozzle-exit boundary-layer conditions on the flow and acoustic fields of initially laminar jets, *Journal of Fluid Mechanics* Vol. 663: pp. 507–538.

Bogey, C., Bailly, C. & Juvé, D. (2003). Noise investigation of a high subsonic, moderate Reynolds number jet using a compressible Large Eddy Simulation, *Theoretical and Computational Fluid Dynamics* Vol. 16: pp. 273–297.

Boris, J. P., Grinstein, F. F., Gran, E. S. & Kolbe, R. L. (1992). New insights into Large Eddy Simulation, *Fluid Dynamics Research* Vol. 10: pp. 199–228.

Bridges, J. & Brown, C. A. (2004). Parametric testing of chevrons on single flow hot jets, *Proceedings of the 10th AIAA/CEAS Aeroacoustics Conference*, number AIAA 2004-2824.

Caruana, D., Barricau, P., Hardy, P., Cambronne, J.-P. & Belinger, A. (2009). The "plasma synthetic jet" actuator. Aero-thermodynamic characterization and first flow control applications, *Proceedings of the 47th AIAA Aerospace Sciences Meeting*, number AIAA 2009-1307.

Castelain, T. (2006). *Contrôle de jet par microjets impactants. Mesure de bruit rayonné et analyse aérodynamique*, PhD thesis, École Centrale de Lyon.

Castelain, T., Sunyach, M., Juvé, D. & Béra, J.-C. (2007). Jet noise reduction by impinging microjets: an aerodynamic investigation testing microjet parameters, *Proceedings of the 13th AIAA/CEAS Aeroacoustics Conference*, number AIAA 2007-3419.

Castelain, T., Sunyach, M., Juvé, D. & Béra, J.-C. (2008). Jet noise reduction by impinging microjets: an acoustic investigation testing microjet parameters, *AIAA Journal* Vol. 46(No. 5): pp. 1081–1087.

Courbet, B., Benoit, C., Couaillier, V., Haider, F., Le Pape, M.-C. & Péron, S. (2011). Space discretization methods, *Aerospace Lab* Vol. 2: pp. 1–14.

Dorey, L.-H., Tessé, L., Bertier, N. & Dupoirieux, F. (2010). A strategy for modeling soot formation and radiative transfer in turbulent flames, *Proceedings of the 14th International Heat Transfer Conference*, number IHTC 14-22819.

Dupoirieux, F. & Bertier, N. (2011). The models of turbulent combustion in the CHARME solver of CEDRE, *Aerospace Lab* Vol. 2: pp. 1–7.

Enomoto, S., Yamamoto, K., Yamashita, K., Tanaka, N., Oba, Y. & Oishi, T. (2011). Large-Eddy Simulation of high-subsonic jet flow with microjet injection, *Proceedings of the 17th AIAA/CEAS Aeroacoustics Conference*, number AIAA 2011-2883.

Eschricht, D., Panek, L., Yan, J., Michel, U. & Thiele, F. (2008). Jet noise prediction of a serrated nozzle, *Proceedings of the 14th AIAA/CEAS Aeroacoustics Conference*, number AIAA 2008-2971.

Ffowcs Williams, J. E. & Hawkings, D. L. (1969). Sound generation by turbulence and surfaces in arbitrary motion, *Philosophical Transactions for the Royal Society of London* Vol. A264(No. 1151): pp. 321–342.

Fisher, M. J., Harper-Bourne, M. & Glegg, S. A. L. (1977). Jet engine noise source location: the polar correlation technique, *Journal of Sound and Vibration* Vol. 51(No. 1): pp. 23–54.

Fleury, V., Bailly, C., Jondeau, E., Michard, M. & Juvé, D. (2008). Space-time correlations in two subsonic jets using dual-PIV measurements, *AIAA Journal* Vol. 46(No. 10): pp. 2498–2509.

Freund, J. B. (2001). Noise sources in a low-Reynolds-number turbulent jet at Mach 0.9, *Journal of Fluid Mechanics* Vol. 438: pp. 277–305.

Fureby, C. & Grinstein, F. F. (1999). Monotonically integrated Large Eddy Simulation of free shear flows, *AIAA Journal* Vol. 37(No. 5): pp. 544–556.

Grinstein, F. F. & Fureby, C. (2002). Recent progress on MILES for high Reynolds-number flows, *Journal of Fluids Engineering* Vol. 124(No. 4): pp. 848–861.

Guillou, F. & Chedevergne, F. (2011). Internal turbine blade cooling simulation: advanced models assesment on ribbed configurations, *Proceedings of the 8th Thermal Engineering Joint Conference*, number AJTEC 2011-44295.

Hardy, P., Barricau, P., Belinger, A., Caruana, D., Cambronne, J.-P. & Gleyzes, C. (2010). Plasma synthetic jet for flow control, *Proceedings of the 40th Fluid Dynamics Conference and Exhibit*, number AIAA 2010-5103.

Huet, M., Fayard, B., Rahier, G. & Vuillot, F. (2009). Numerical investigation of the micro-jets efficiency for jet noise reduction, *Proceedings of the 15th AIAA/CEAS Aeroacoustics Conference*, number AIAA 2009-3127.

Husain, Z. D. & Hussain, A. K. M. F. (1979). Axisymmetric mixing layer: Influence of the initial and boundary conditions, *AIAA Journal* Vol. 17: pp. 48–55.

Kastner, J., Kim, J.-H. & Samimy, M. (2008). Toward better understanding of far-field radiated noise mechanisms in a high Reynolds number Mach 0.9 axisymmetric jet, *Proceedings of the 46th AIAA Aerospace Sciences Meeting and Exhibit*, number AIAA 2008-7.

Kearney-Fischer, M., Kim, J.-H. & Samimy, M. (2009a). Control of a high Reynolds number Mach 0.9 heated jet using plasma actuators, *Physics of Fluids* Vol. 21: 095101.

Kearney-Fischer, M., Kim, J.-H. & Samimy, M. (2009b). Noise control of a high Reynolds number Mach 0.9 heated jet using plasma actuators, *Proceedings of the 15th AIAA/CEAS Aeroacoustics Conference*, number AIAA 2009-3188.

Kearney-Fischer, M., Kim, J.-H. & Samimy, M. (2011). A study of Mach wave radiation in an axisymmetric jet using active control, *Proceedings of the 17th AIAA/CEAS Aeroacoustics Conference*, number AIAA 2011-2834.

Kim, J.-H., Kastner, J. & Samimy, M. (2009). Active control of a high Reynolds number Mach 0.9 axisymmetric jet, *AIAA Journal* Vol. 47(No. 1): pp. 116–128.

Krothapalli, A., Venkatakrishnan, L., Lourenco, L., Greska, B. & Elavarasan, R. (2003). Turbulence and noise suppression of a high-speed jet by water injection, *Journal of Fluid Mechanics* Vol. 491: pp. 131–159.

Laufer, J., Schlinker, R. & Kaplan, R. E. (1976). Experiments on supersonic jet noise, *AIAA Journal* Vol. 14(No. 4): pp. 489–497.

Laurendeau, E., Jordan, P., Bonnet, J.-P., Delville, J., Parnaudeau, P. & Lamballais, E. (2008). Subsonic jet noise reduction by fluidic control: The interaction region and the global effect, *Physics of Fluids* Vol. 20: 101519.

Lew, P.-T., A.Najafi-Yazdi & L.Mongeau (2010). Unsteady numerical simulation of a round jet with impinging microjets for noise suppression, *Proceedings of the 48th AIAA Aerospace Sciences Meeting*, number AIAA 2010-18.

Lighthill, M. J. (1952). On sound generated aerodynamically, part I: General theory, *Proceedings of the Royal Society of London* Vol. A221(No. 1107): pp. 564–587.

Lyrintzis, A. S. (1994). Review: the use of Kirchhoff's method in computational aeroacoustics, *Journal of Fluids Engineering* Vol.(No. 4): pp. 665–676.

Michalke, A. (1965). On spatially growing disturbances in an inviscid shear layer, *Journal of Fluid Mechanics* Vol. 23: pp. 521–544.

Muller, F. (2006). *Simulation de jets propulsifs : application à l'identification des mécanismes générateurs de bruit*, PhD thesis, Université Paris VI.

Muller, F., Vuillot, F., Rahier, G., Casalis, G. & Piot, E. (2006). Experimental and numerical investigation of the near field pressure of a high subsonic hot jet, *Proceedings of the 12th AIAA/CEAS Aeroacoustics Conference*, number AIAA 2006-2535.

Najafi-Yazdi, A., Lew, P.-T. & Mongeau, L. (2011). Large Eddy Simulation of jet noise suppression by impinging microjets, *Proceedings of the 17th AIAA/CEAS Aeroacoustics Conference*, number AIAA 2011-2748.

Nesbitt, E. & Young, R. (2008). Forward flight effects on chevron noise reduction, *Proceedings of the 14th AIAA/CEAS Aeroacoustics Conference*, number AIAA 2008-3065.

Perez, G., Prieur, J., Rahier, G. & Vuillot, F. (2007). Theoretical and numerical discussion on volume integral methods for jet noise prediction, *Proceedings of the 13th AIAA/CEAS Aeroacoustics Conference*, number AIAA 2007-3593.

Ragaller, P. A., Annaswamy, A. M., Greska, B. & Krothapalli, A. (2009). Supersonic jet noise reduction via pulsed microjet injection, *Proceedings of the 15th AIAA/CEAS Aeroacoustics Conference*, number AIAA 2009-3224.

Rahier, G., Prieur, J., Vuillot, F., Lupoglazoff, N. & Biancherin, A. (2004). Investigation of integral surface formulations for acoustic post-processing of unsteady aerodynamic jet simulations, *Aerospace Science and Technology* Vol. 8: pp. 453–467.

Rife, M. E. & Page, G. J. (2011). Large Eddy Simulation of high Reynolds number jets with microjet injection, *Proceedings of the 17th AIAA/CEAS Aeroacoustics Conference*, number AIAA 2011-2882.

Samimy, M., Kim, J.-H., Kastner, J., Adamovich, I. & Utkin, Y. (2007a). Active control of a Mach 0.9 jet for noise mitigation using plasma actuators, *AIAA Journal* Vol. 45(No. 4): pp. 890–901.

Samimy, M., Kim, J.-H., Kastner, J., Adamovich, I. & Utkin, Y. (2007b). Active control of high-speed and high-Reynolds-number jets using plasma actuators, *Journal of Fluid Mechanics* Vol. 578: pp. 305–330.

Shur, M. L., Spalart, P. R. & Strelets, M. K. (2011). LES-based evaluation of a microjet noise reduction concept in static and flight conditions, *Journal of Sound and Vibration* Vol. 330: pp. 4083–4097.

Smagorinsky, J. (1963). General circulation experiments with the primitive equations, *Monthly Weather Review* Vol. 91(No. 3): pp. 99–164.

Tam, C. K. W., Viswanathan, K., Ahuja, K. K. & Panda, J. (2008). The sources of jet noise: experimental evidence, *Journal of Fluid Mechanics* Vol. 615: pp. 253–292.

Tanna, H. (1977). An experimental study of jet noise, part I: Turbulent mixing noise; part II: Shock associated noise, *Journal of Sound and Vibration* Vol. 50(No. 3): pp. 405–444.

Tanna, H., Dean, P. & Fisher, M. (1975). The influence of the temperature on shock-free supersonic jet noise, *Journal of Sound and Vibration* Vol. 39(No. 4): pp. 429–460.

Uzun, A. & Hussaini, M. Y. (2009). Simulation of noise generation in near-nozzle region of a chevron nozzle jet, *AIAA Journal* Vol. 47(No. 8): pp. 1793–1810.

Viswanathan, K. (2004). Aeroacoustics of hot jets, *Journal of Fluid Mechanics* Vol. 516: pp. 39–82.

Vuillot, F., Lupoglazoff, N. & Huet, M. (2011). Effect of chevrons on double stream jet noise from hybrid CAA computations, *Proceedings of the 49th AIAA Aerospace Sciences Meeting*, number AIAA 2011-1154.

Xia, H., Tucker, P. G. & Eastwood, S. (2008). Towards jet flow LES of conceptual nozzles for acoustic predictions, *Proceedings of the 46th AIAA Aerospace Sciences Meeting and Exhibit*, number AIAA 2008-10.

Yan, J., Tawackolian, K., Michel, U. & Thiele, F. (2007). Computation of jet noise using a hybrid approach, *Proceedings of the 13th AIAA/CEAS Aeroacoustics Conference*, number AIAA 2007-3621.

Zaman, K. B. M. Q. (1985). Far-field noise of a subsonic jet under controlled excitation, *Journal of Fluid Mechanics* Vol. 152: pp. 83–111.

Zaman, K. B. M. Q. (2010). Subsonic jet noise reduction by microjets - a parametric study, *International Journal of Aeroacoustics* Vol. 9(No. 6): pp. 705–732.

5

Comparison of Aerodynamic Effects Promoted by Mechanical and Fluidic Miniflaps on an Airfoil NACA 4412

M. Casper[1], P. Scholz[1], J. Colman[2],
J. Marañón Di Leo[2], S. Delnero[2] and M. Camocardi[2]
[1]Institute of Fluid Mechanics, Braunschweig Technical University
[2]Boundary Layer and Environmental Fluid Dynamics Laboratory,
Engineering Faculty, National University of La Plata
[1]Germany
[2]Argentina

1. Introduction

Miniflaps, often called "Gurney flaps" have been introduced by the US race-car driver Dan Gurney, who used these small-scale (typically 1% of chord), highly deployed flaps at the trailing edge of the rear spoiler of his racing car. He experienced a significant increase of lift.

Since then Gurney flap were studied by many authors [2]- [5], all confirming an increase of the lift and lift-to-drag ratio and a reduced form drag obtained at high lift coefficients compared with the same airfoil without the flap. Giguére et al [6], described experimentally the aerodynamic behavior of these flaps scaling their height with the boundary layer thickness.

Although Gurneys are extensively used in race-car aerodynamics, application on aircraft is rather uncommon. That might be attributed to the fact that it is very intricate to design deployable flaps in the very thin trailing edge region of common airfoils and the effectiveness of Gurney flaps does not justify the effort or additional weight, respectively. Therefore it is of great interest to explore fundamental methods to increase the effectiveness.

The wake immediately downstream a common lifting airfoil is asymmetric due to the different external and boundary layer flows on the pressure and suction surfaces of the airfoil. The structure in this near wake region is influenced by aero dynamical loading and airfoil characteristics.

Downstream of the trailing edge of a normal lifting airfoil the downwash diminishes rapidly. Hah and Lakshminarayana, in their experimental and numerical study about the near wake of a lifting airfoil [1], confirmed that the asymmetric wake becomes nearly symmetric after only one chord length downstream of the trailing edge. These authors reported also that the far wake shows a roughly symmetric wake structure in which the airfoil features and aerodynamic loads does not influence the wake development anymore.

Also, the fundamental understanding of the lift increase generated by Gurney-type miniflaps is connected to a wake phenomenon: A counter-rotating vortex pair will develop behind the blunt trailing-edge generated by the Gurney flap. These vortices induce streamlines resembling a smooth, "virtual" aerodynamic prolongation of the airfoil, allowing to retain a pressure difference at the trailing edge and thus adding a virtual camber by shifting downward the rear stagnation point where the Kutta condition must be applied. A more detailed description of the flow structures in the Gurney flap flow-pattern obtained by experiments with laser Doppler anemometry (LDA) was reported [7], [8]. Aspects of the behaviour of airfoils equipped with miniflaps with different lengths were described in [9], reported about the influence of free stream turbulence structure on these devices.

However, the unsteady nature of the counter-rotating vortices behind the miniflap complicates this "time-averaged" point of thinking. It is known that the highly unsteady nature of flows associated with vortex shedding make theoretical understanding difficult. It should be mentioned that up to date there are no theories for predicting the drag coefficient as a function of Reynolds number in vortex shedding conditions of extremely basic bodies such as circular cylinders.

In some studies about aerodynamic efficiency of Gurney flaps it is common to find the assumption that the effect of the asymmetry of the vortices on lift to drag ratio is detrimental. In order to attain a drag reduction by "*stabilizing the wake*" some authors suggest the use of span-wise holes, slits, serrated flaps and wake-bodies [10], [11], while other recommend to eliminate straightforward the double row of counter-rotating vortices of the wake behind a Gurney type miniflap [12].

2. Our proposal

We hypothesize that different perturbations introduced near the separating regions of the shear layers from the pressure and suction sides of an airfoil introduce asymmetry in the vortex structures of the near wake region able to produce lift enhancing effects.

The turbulent flow behind the blunt trailing edge of an airfoil with a Gurney flap, involves turbulent boundary layers separating at the airfoil edges, developing into shear layers which rollup into discrete vortices. The negative and positive vorticity from the separating boundary layers is packed into the rolling up shear layers developing into vortices that merge downstream at the vortex shedding frequency. The Kutta condition is obviously not fulfilled at the blunt trailing edge, but at a moving stagnation point that appears rearward and downward from the trailing edge of the airfoil. Each time a vortex is shed into the wake circulation is produced. According Kelvin's theorem, the periodic vortex shedding from the trailing edge into the wake is connected to time variations of the strength of the bound vortex, which in turn deviate the velocity field, inducing periodic variations of the angle of attack of the incident velocity. Let us concentrate on the near wake flow behind blunt bodies. A periodic shedding of equal counter rotating vortices creates a periodic up and downward lift and periodic drag variations. A body with a Karman type vortex wake, experiences a periodic fluctuating upward and downward lift coupled to the regular counter rotating vortex shedding which induces upward and downward deflections of the flow. Hence, a symmetric Karman type vortex street pattern alone could not contribute to a steady lift.

The lift required in common aeronautic applications involves a preferential deviation of the oncoming flow visualized by the known up and downwash before and behind a wing. If we focus on the downwash of the flow behind the trailing edge of a conventional lifting airfoil, we appreciate that more lift is connected with more downwash. Therefore, a vortex street in itself is not detrimental, but asymmetry of the wake is crucial. E.g. in the near wake region the initial eddies of the vortex street rotating in one direction, should be much stronger than the initial eddies rotating in the opposite direction such that a preferential deflection of the wake connected to a time averaged steady lift will occur.

Therefore it is worthwhile to explore the vertical motion and dynamics in the near-wake of an airfoil with a Gurney-type miniflap. It is even more worthwhile to introduce a mechanism that will be able to actively influence and maximize the asymmetry of the wake in order to increase the additional lift gained by the miniflap. Such mechanism can be either a mechanically moving Gurney or a fluidic device that can be switched off and on by the means of valves.

3. Experimental procedure

In the following works, the experiments with the wing model, equipped inside with a electromechanical system which bring oscillating up-down movement to a miniflap, were carried out at the Boundary Layer and Environmental Fluid Dynamics Laboratory (LaCLyFA) closed circuit wind tunnel, at the Faculty of Engineering, National University of La Plata, Argentina. In each work there will be indicated the specific Reynolds number but, in all cases, corresponds to *low Reynolds number aerodynamics*. At LaCLyFA, instantaneous velocities at the near wake were measured by a six channel Dantec Streamline hot-wire constant temperature anemometer (X-wire type probes).

Complementary to these tests with a mechanical gurney experiments, another ones using a fluidic gurney are conducted in the low speed facility at the Institute of Fluid Mechanics (ISM) at Braunschweig Technical University. The model tested, with the NACA 4412 airfoil, has a chord of 240mm. Seven fast switching valves are integrated into the model, the air is guided from the vales through individual "diffusers" into a small settling chamber and then through a small, wall-normal slit into the flow. The position of the slit is x/c=0,92, that is 8% upstream of the trailing edge. Thus, the position of this fluidic gurney is similar to the mechanical gurney at the LaCLyFA. The air emitted through the slit acts just like a jet-flap or a Gurney flap, respectively. Through the valves the effect can be switched on and off, where the valves are small electro-magnetic devices (FESTO MHJ), that are able to produce rather high frequencies of up to 1 kHz in a very reproducible manner. The airfoil is equipped with 30 pressure taps to measure the pressure distribution.

4. Experimental results

4.1 Fluidic mini-flap

Tests with a fluidic Gurney are conducted in the low speed facility at the Institut für Strömungsmechanik (ISM) at Technische Universität in Braunschweig, Germany. The facility is an atmospheric open-jet return-tunnel with a jet diameter of 0.5 m. The jet collector is located 0.98 m downstream of the nozzle. The tunnel has a frequency controlled drive system with 22 kW, enabling a maximum velocity of 60m/s. For the measurements herein however, velocities of 26 m/s and 39 m/s were used to establish Reynolds-identity with the data taken at LaCLyFA. The tunnel velocity is controlled by the differential pressure over the nozzle.

The airfoil is mounted with endplates in the center of the wind tunnel jet. The airfoil model has a chord of c=230 mm, it is equipped with 30 static pressure taps near the center section. To create the fluidic gurney-jet six solenoid fast-switching valves are integrated into the airfoil: the air is guided from the vales through individual "diffusers" into a small settling chamber and then through a small, wall-normal slit into the flow. The position of the slit is x/c = 0.92, that is 8% upstream of the trailing edge. Thus, the position of this fluidic gurney is identical to the mechanical gurney at the LaCLyFA. The slit itself is manufactured using a laser cutting technique into an "insert-sheet" made of stainless steel. The slit size is 0.2mm, because of the manufactured technique the slit size can be controlled very accurately. Through the valves the effect can be switched on and off, where the valves are small electro-magnetic devices (FESTO MHJ) that are able to produce rather high frequencies of up to 1 kHz in a very reproducible manner. The valves are controlled by a square signal with the desired frequency synthesized by a standard frequency generator.

Figure 1a shows a sketch drawing highlighting the fluidic gurney-system. The airfoil body is manufactured using a 3D-printing technique ("Stereo lithography"), where all air ducts, pressure taps and settling chambers are printed in one step. To get access to the valves located inside the airfoil, the body features two large hatches on the pressure side that can be closed with covers representing the airfoil geometry (see also Figure 1b).

The static pressure distribution was measured by means of a custom build multi-channel pressure scanner, based on sensors with a total pressure range of ±250 mbar. Lift can be determined by integration of the static pressure distribution.

The data was corrected for open-jet effects following the procedure described in [13]. The relevant factors are G0=1.0, G1=0.33 for the streamline-curvature correction and ε_S=-0.00867 for solid blockage. Following that procedure the angle of attack that will be compared below (specifically 5°) represents the same aerodynamic condition, although the geometrical angle of attack of the airfoil in the wind tunnel stream is 10°.

Flow analysis was conducted using a PIV setup, consisting of a Litron Nano-T 200mJ/Puls double-pulse Laser and a LaVision Imager Pro X 11M camera with 11 Mega-Pixels. The PIV data has been acquired, correlated and analyzed using the "DaVis 7" software package from LaVision. Correlation was done with an iterative multi-grid procedure with initial window size of 128 Pxs² and final size of 64 Pxs², both with 50% window overlap. Final resolution of the vector field is 2.2 mm. In the case of a pulsing Gurney jet the acquisition trigger of the PIV-system was controlled by the frequency generator of the valves, enabling phase-locked measurements. Due to the sidewall-plates surrounding the airfoil optical access is limited to the region shown in the figures below.

Data has been acquired for a reference case (no fluidic Gurney jet), a statically blowing Gurney jet (comparable to a "classical" mechanical Gurney that is deployed stationary) and a pulsing Gurney jet. Pulsing frequency is F1 (f⁺=1.32). For the pulsing jet three different phase positions were measured, namely λ=0.13 ·T (or 45°), λ=0.49 ·T (or 180°) and λ=0.77 ·T (or 275°), where T is the periodicity time of the pulsing.

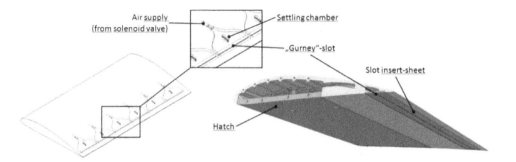

Fig. 1a. Sketch of the fluidic-gurney airfoil showing the air supply system

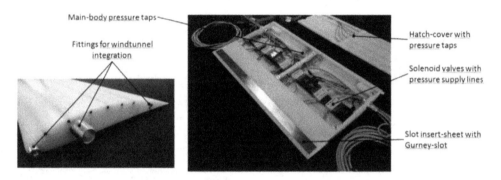

Fig. 1b. Photographs of the airfoil model with fluidic mini-flap system

Results (medium to far wake region)

The following figures highlight the vorticity field that was computed from the velocities. The vorticity field was averaged over 500 individual measurements for each of the cases. Vorticity is in an arbitrary unit (based on the image scaling), which is consistent over all measurements.

Figure 2 to 5 highlights the transversal vorticity, normalized by the peak vorticity measured for the dynamic gurney jet, in comparison for the reference case (without Gurney jet), the statically blowing Gurney jet and the pulsed Gurney. The pulsing frequency is again F1 (reduced frequency $f^+=1.3$). The time averaged flow rate for the pulsed Gurney and the static Gurney is the same. Solid lines and/or open symbols represent positive vorticity, dashed lines and/or filled symbols represent negative vorticity. Note that for the two cases reference and static Gurney the vorticity over x/c is essentially a continuous curve, since the data is averaged over time. For the pulsed Gurney the data points represent the peak vorticity (and respective position) of individual vortices that have rolled up in the shear layer. These vortices have been manually isolated piece by piece, noting vorticity and position. This analysis has been done with different phase angles (see Figure 6), thus the data represents the averaged peak vorticity of local vortices. Furthermore it must be noted that positive vorticity is "rotation counterclockwise", thus positive vorticity should be connected to a lift increase at the airfoil, while negative (rotation clockwise) vorticity to lift decrease.

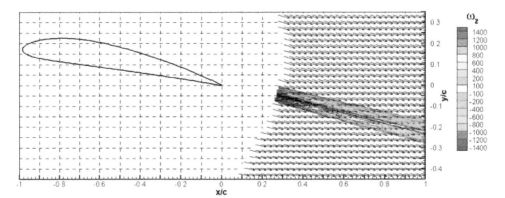

Fig. 2. Vector-field and transversal vorticity (every 3rd vector displayed) for the reference case (no Gurney)

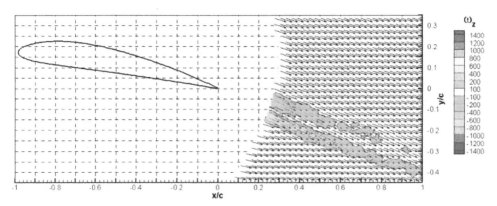

Fig. 3. Same as above for the statically blowing fluidic Gurney jet

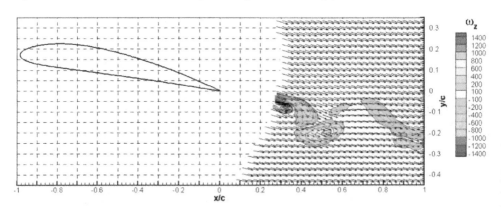

Fig. 4. Same as above for the pulsed fluidic Gurney jet, pulsing $f^+=1.32$

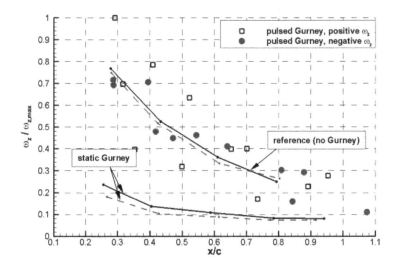

Fig. 5. Transversal vorticity over downstream distance for the different cases

Some different conclusions can be drawn from the plot: For the reference case the wake is almost symmetric, with however, consistently slightly weaker negative vorticity. The same is true for the statically blowing Gurney jet, where the total level of vorticity is significant lower, but the asymmetry is slightly more pronounced compared to the reference case.

The individual vortices emanating from the pulsed Gurney jet are quite strong compared to the two other cases. Although is evident that with only three phase angles it is not possible to draw a continuous line for the peak vorticity of the vortices carried out downstream, most of the vortices rotating counterclockwise (positive) are stronger than their negative counterparts. This is especially true the smaller x/c is (e.g. the closer we get to the near wake region), while respectively with increasing x/c also the difference in positive and negative vorticity vanishes and the vortices become more and more symmetric.

Following the above stated hypothesis this is due to the lift that the airfoil creates. For all cases the wake asymmetry vanishes somewhat around $x/c=1$, that is one chord length downstream of the airfoil.

The Figure 6 highlights the temporal evolution of the vorticity in a spatially resolved, closeup view of the wake. It is clearly visible at time 0.13·T that a almost isolated, rather strong vortex develops at the trailing edge, when the Gurney jet is turned on. This vortex rotates counterclockwise and thus contributes to lift or is connected to a positive lift increase at the airfoil, respectively. This vortex trails downstream and decays slowly. In the later stages a quasi-static wake develops, which is quite similar to the wake without Gurney, since this is connected to the part of the period where the Gurney jet is switched off. Note also that the most pronounced peak negative vorticity occurs more in this quasi-static part of the period, whereas the peak positive vorticity is connected to the "puff" that comes from the suddenly starting Gurney.

λ=0.13 ·T:

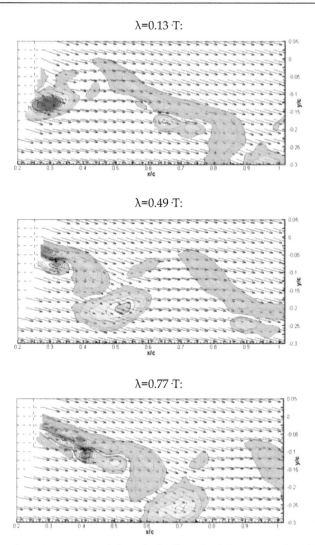

λ=0.49 ·T:

λ=0.77 ·T:

Fig. 6. Evolution of vorticity in time (phase locked), pulsed Gurney jet f⁺=1.3; vorticity color same as in Figure 2

4.2 Mechanical mini-flap

The untwisted wing model tested, of 60cm chord (c) and 80cm wingspan had a NACA 4412 airfoil. The Reynolds number, based upon the wing chord and the mean free stream velocity were 326.000 and 489000, based upon the mean free stream velocity (at 1.5m ahead the model at its height) and the model chord, corresponding to values of it of 10m/s and 15m/s respectively, depending on the test. The turbulence intensity was 1.8% (minimum turbulence intensity of the wind tunnel). The closest possible distance from the trailing edge, for the Gurney flap position, was determined by the implemented mechanism inside the model.

The model has a mini-flap Gurney type located on the lower surface near the trailing edge, used as passive and active flow control device. As an active device an up-down movement was designed, geared by an electromechanical system. Such mechanism employed a continuous current small electric motor with a crankshaft fixed at its axle, provided with a high capacity rotation velocity sealed bearing. The bearing, when the motor rotate, impinge an up-down movement to the mini-flap (composite plate) placed along wingspan at 8%c from the trailing edge. Mini-flap height was 1.3%c. Thus, the mini-flap was capable to make an oscillatory up-down movement from the lower surface (0mm displacement) to the maximum amplitude (10mm). The electric motor had a constant voltage variable frequency power supply, in order to be able to vary the up-down movement frequency. Figure 7 shows a schematic draw of the model with the electromechanical system inside it:

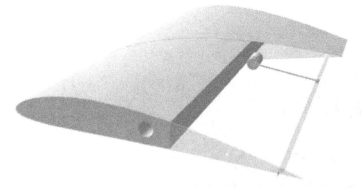

Fig. 7. Schematic draw of the model wing with the electromechanical mechanism

The experiments were carried out in the closed circuit boundary layer wind tunnel of the Boundary Layer & Environmental Fluid Dynamics Laboratory at the Aeronautical Department, Engineering Faculty, National University of La Plata (Argentina). The test section is 1.4m wide, 1m height and 7.5m length and has upstream flow deflectors in order to achieve the desired mean free stream velocity and turbulence intensity.

Figure 8 shows a diagram of the wing model inside the wind tunnel test section. In Figure 9 we show views of the experimental setup, the wing model was mounted between two double lateral plates, which trailing edges were capable to manually adjust along each axis, in order to have a favorable (or almost zero) pressure gradient along the test section, with the thinnest possible boundary layer

Figure 10 shows a schema of the wake measuring positions, along "x" axis and the corresponding vertical points (1H, 2H and 5H), indicating in this Figure only the 0 and -10 points, the separation between vertical adjacent points is 0.25H (2 mm).

A schema of the electromechanical system is shown in Figure 11.

The objective of the experiments carried out at LaCLyFA, were to measure instantaneous velocities in the near wake region, because the near wake structure is directly related with lift production (see also above explanations). Designing with "x" the horizontal coordinate (which is coincident with the model chord line for 0^0 degree angle of attack) and with "y"

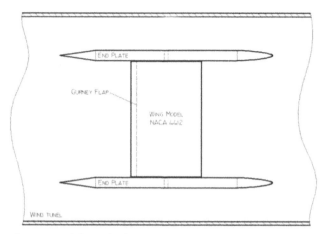

Fig. 8. Diagram of wing model inside the wind tunnel test section

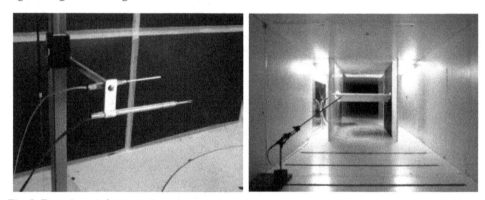

Fig. 9. Experimental setup views.

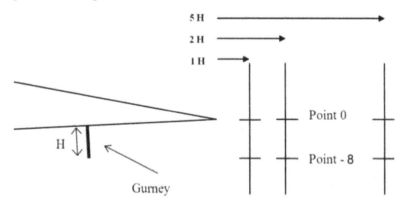

Fig. 10. Measurements Grid

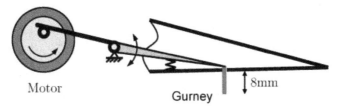

Fig. 11. Mini-flap Gurney type electromechanical movement (schematic).

the vertical coordinate, velocity measurements were performed at 1H, 2H and 5H "x" positions in the wake, being "H" the mini-flap height. For each "x" position, vertical measurements were performed from y = 1H to y = - 2H with intervals of 0.05H, being y = 0 the trailing edge level.

Velocity data in the near wake, as mentioned above, are very important in order to qualify and quantify the vortex characteristics and structure, together with turbulence intensities, because the possible lift increment is closed related with such near wake structure, that's, if is strongly asymmetric or not.

Instantaneous velocities were measured with a hot wire constant temperature anemometer Dantec Streamline, using X-probes. The acquisition frequency was 4000 Hz per channel, filtered at 2000 Hz.

The acquired data were very extensive and, for such reason, both research teams (Germany and Argentine teams) for the comparison objectives cited in this work choose results for a determined angle of attack, say 5^0 degrees.

5. Results analysis (near and medium wake region)

Figure 12 shows the mean velocities U and V, plotted for the five situations-without mini-flap; with fixed deployed mini-flap; oscillating mini-flap for three frequencies, 22Hz (Frequency 1), 38Hz (Frequency 2) and 44Hz (Frequency 3), all for the three "x" positions.

Regarding the Figure 12, we could conclude that the U-component has small variations between the different conditions (clean airfoil; fixed flap; etc), being always positive above and below the trailing edge, but with a reduction of its magnitude from the trailing edge level to the end of mini-flap level. The vertical velocities exhibit important differences, above the trailing edge, between the clean airfoil and the fixed mini-flap case. Respect the oscillating mini-flap, there are small differences between the vertical velocities for the three frequencies but, if we look close the vertical velocities at the mini-flap level and lower, their values are greater than the corresponding to clean airfoil or even the fixed mini-flap case. Qualitatively, the situation is similar for all the position. Analyzing the Figure, it's clear that we have an anticlockwise vortex behind the mini-flap. This is consistent with the results founded by other authors.

Figure 13 shows mean vorticity distribution, along "x" coordinate", evaluated as 2D with the well known expression $\partial V/\partial x - \partial U/\partial y$. In this Figure we show that there is a decrement in the vorticity field downstream the trailing edge, corresponding to the diffusion of the vortex generated by the Gurney flap. Note that the conclusions from the instantaneous

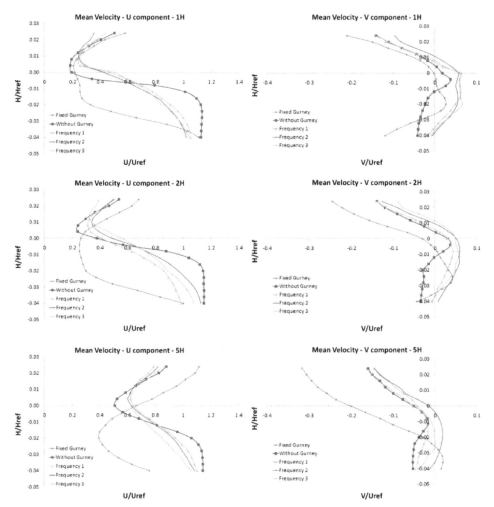

Fig. 12. Mean velocity distributions for the different positions and all the cases.

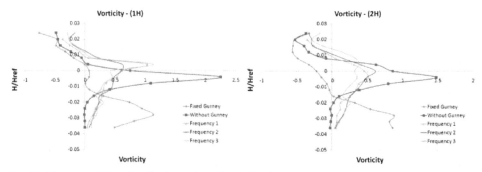

Fig. 13. Mean vorticity distribution along the "x" axis.

vorticity plot of the fluidic Gurney (Figures 2, 3 and 4) hold here as well and are comparable with the mean vorticity calculations showed for the mechanical mini-flap case. The peak positive vorticity in the lower wake is more pronounced than the (negative) vorticity in the upper wake. This is true for all cases.

Figure 14 corresponds to turbulence intensities for the three "x" positions. Again the flow turbulent intensities diminish as we move downstream the trailing edge, showing to us that the vortex generated by the Gurney seems to disappear after 5H position.

Regarding Figure 15, almost all positions didn´t revealed the spectra peak from the Gurney vortex downstream, except for the movable one that show us a series of peaks referring the activity of the flap. Also, the peaks found in some of them are related to the frequencies of

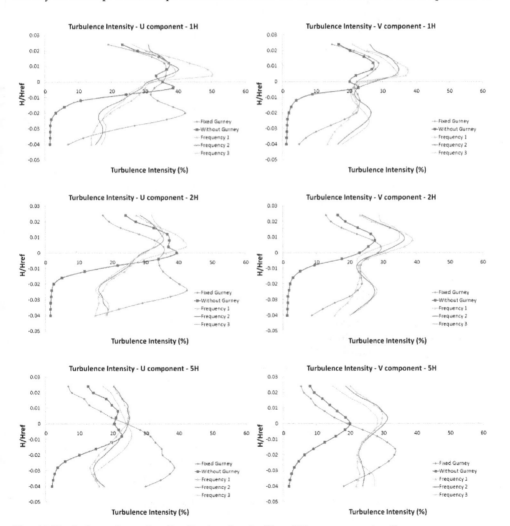

Fig. 14. Turbulence intensity distribution for the U and V component in all cases

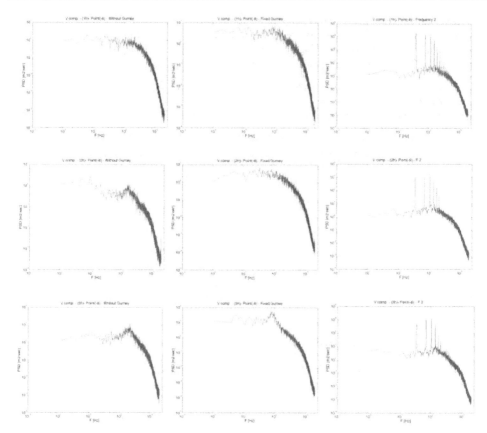

Fig. 15. Power Density Spectra at point -8, with and without Gurney Flap, and the movable one at frequency 2. In the location at 1, 2 and 5 H positions, downstream the trailing edge airfoil at 5° angle of attack.

the vortex release by the airfoil and not with those generated by the fixed Gurney (200 Hz approximately). This means that the Gurney downstream is outside the measured grid, except for the case of 5H at point -8 for the fixed Gurney, in which the 96 Hz peak show to us the presence of the Gurney vortex downstream. From this behavior we can conclude, for this angle of attack, that the vortex generated by the Gurney flap it is moving downstream and will be catch a long distances from the trailing edge before it becomes diffuse at the far wake.

Figures 16, 17 and 18 corresponds to the autocorrelation coefficients, at 1H, for horizontal and vertical instantaneous velocities, for the case with fixed miniflap and the oscillatory one for Frequencies 1 and 3, respectively. Figures 19 to 21, corresponds to the autocorrelations coefficients, at 2H, as previous Figures but for Frequencies 1 and 2. Finally, Figures 22 to 23 referred also to the autocorrelation coefficients, as Figures 16 to 18, but for 5H.

Table 1 also shows temporal and spatial scales for the horizontal and vertical components of the mean velocities in the near wake, for the five situations: without miniflap; fixed miniflap and oscillatory one (three excitation frequencies).

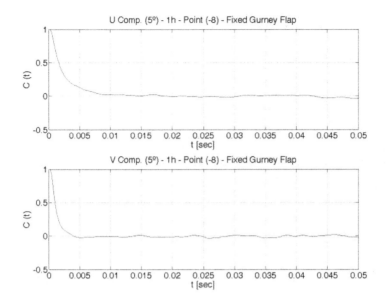

Fig. 16. Autocorrelation coefficients for 1H at fixed Gurney.

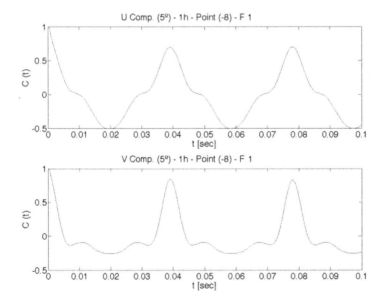

Fig. 17. Autocorrelation coefficients for 1H at Frequency 1.

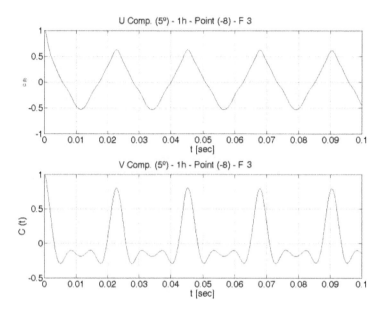

Fig. 18. Autocorrelation coefficients for 1H at Frequency 3.

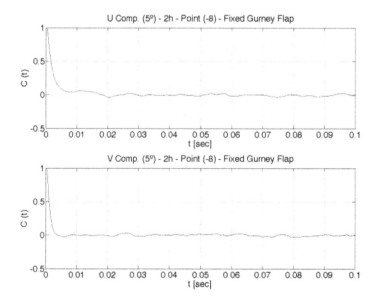

Fig. 19. Autocorrelation coefficients for 2H at fixed Gurney.

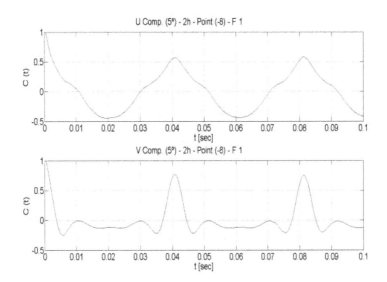

Fig. 20. Autocorrelation coefficients for 2H at Frequency 1.

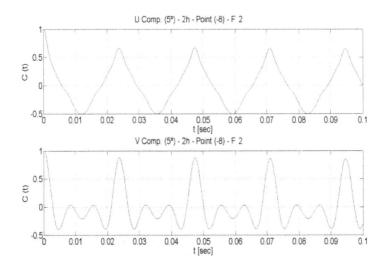

Fig. 21. Autocorrelation coefficients for 2H at Frequency 2.

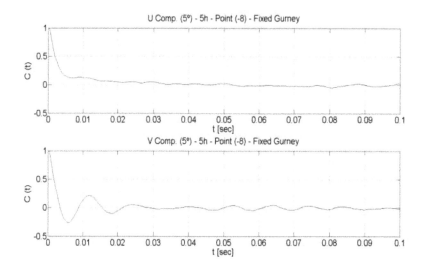

Fig. 22. Autocorrelation coefficients for 5H at fixed Gurney.

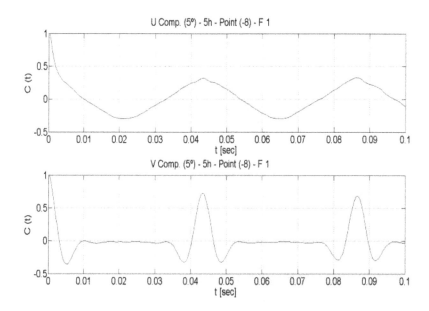

Fig. 23. Autocorrelation coefficients for 5H at Frequency 1.

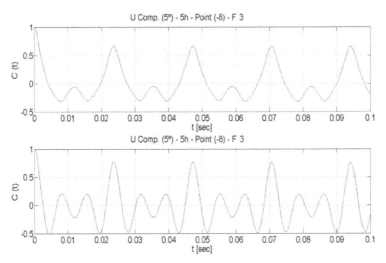

Fig. 24. Autocorrelation coefficients for 5H at Frequency 3.

	Device	tu [sec]	tv [sec]	U [m/sec]	V [m/sec]	Eu [m]	Ev [m]
1H	Without Gurney	0.01350	0.03950	10.955	0.156	0.14789	0.00616
	Fixed Gurney	0.01875	0.00400	2.688	0.236	0.05040	0.00094
	Frequency 1	0.00975	0.00500	8.454	0.615	0.08243	0.00308
	Frequency 2	0.00675	0.00325	8.178	0.557	0.05520	0.00181
	Frequency 3	0.00575	0.00325	8.025	0.623	0.04614	0.00202

	Device	tu [sec]	tv [sec]	U [m/sec]	V [m/sec]	Eu [m]	Ev [m]
2H	Without Gurney	0.01275	0.00200	11.210	0.152	0.14293	0.00030
	Fixed Gurney	0.01800	0.00400	2.780	0.236	0.05004	0.00094
	Frequency 1	0.01075	0.00375	7.986	0.363	0.08585	0.00136
	Frequency 2	0.00775	0.00300	9.066	0.580	0.07026	0.00174
	Frequency 3	0.00575	0.00275	8.489	0.500	0.04881	0.00138

	Device	tu [sec]	tv [sec]	U [m/sec]	V [m/sec]	Eu [m]	Ev [m]
5H	Without Gurney	0.01475	0.00150	9.590	0.271	0.14145	0.00041
	Fixed Gurney	0.05275	0.00350	3.987	0.615	0.21031	0.00215
	Frequency 1	0.01000	0.00325	8.940	0.170	0.08940	0.00055
	Frequency 2	0.00727	0.00300	8.742	0.186	0.06355	0.00056
	Frequency 3	0.00425	0.00250	8.017	0.010	0.03407	0.00003

Table 1. Temporal (tu and tv) and Spatial (Eu and Ev) scales for u and v velocities components.

As a first plain sight we could establish that:

a. Regarding turbulence intensity and spatial scales also, along "x" coordinate, it seems to slow both as miniflap frequency grow. This behavior was observed, also, for the wing model with the mechanical miniflap, at an angle of attack of 0^0. Due space reasons of this article, we don´t show also the results for such angle of attack.

b. Mean vertical velocity V is greater for the moving flap than for the fixed one, all in the near wake region. This effect becomes lower as we go far away the near wake.

c.   Mean vorticity, always in the near wake region, seems to become slow with the fixed mini-flap in comparison with the clean airfoil and, as oscillating frequency increase, such mean vorticity diminishing effect becomes more important than with the fixed mini-flap. The mean vorticity behavior is similar to the results obtained by Wassen et al [14], at the same points measured by us in the present work.

d.   The above cited behavior of the mean vorticity, at the near wake, has a good match with the instantaneous vorticity results provided by fluidic miniflap experiments. Due the fact that the vorticity calculations for the mechanical miniflap are done with the mean velocities, and for the fluidic miniflap were done with the instantaneous velocities, the mentioned similar behavior of both experiments is qualitative but, in that sense, the agreement is good.

e.   As the angle of attack grows, vertical mean velocity becomes more important than horizontal one.

The above summarized analysis encourage us to go deep, in the future, into the study of the effects of fluidic and mechanical miniflaps upon lift behavior and, hence, upon aerodynamic efficiency of airfoils with such active flow control mechanism.

6. References

[1] C. Hah et al, Measurement and prediction of mean velocity and turbulence structure in the near wake of an airfoil, *Journal of. Fluid Mechanics. Vol.115, 1982 pp. 251-282.*

[2] R.H. Liebeck, Design of subsonic airfoils for high lift, *Journal of Aircraft Vol. 15, No. 9, 1978, pp 547-561.*

[3] D.H. Neuhart et al, A water tunnel study of Gurney flaps, *NASA TM-4071, 1988.*

[4] A.W. Bloy, Aerodynamic Characteristics of an aerofoil with Small Trailing Edge Flaps, *Wind Engineering, Vol. 19, No.3, 1995, pp 167-172.*

[5] B.L. Storms, Lift Enhancement of an Airfoil Using a Gurney Flap and Vortex Generators, *Journal of Aircraft Vol. 31, No. 3, 1994, pp 542-547.*

[6] P. Giguére, J. Lemay, G. Dumas, Gurney Flap Effects and Scaling for Low-Speed Airfoils, *AIAA Paper 95-1881, 13th AIAA Applied Aerodynamics Conference San Diego, 1995.*

[7] D.R.M. Jeffery, D.W. Hurst, Aerodynamics of the Gurney Flap, *AIAA Applied Aerodynamic Conference, AIAA 96-2418-CP, 1996.*

[8] D. Jeffrey et al, Aerodynamics of Gurney Flaps on a Single-Element High-Lift Wing, *Journal of Aircraft, Vol. 37, 2000, pp. 295-301.*

[9] F. Bacchi, J. Marañón Di Leo, J.S. Delnero, J. Colman, M. Martinez, M. Camocardi, U. Boldes, Determinación experimental del efecto de mini flaps Gurney sobre un perfil HQ–17, *Fluidos-2006 IX Reunión Sobre Recientes Avances En Física de Fluidos y sus Aplicaciones, Mendoza, Argentina, 2006.*

[10] C.P. Van Dam et al, Gurney Flap Experiments on Airfoil and Wings, *Journal of Aircraft (0021-8669), Vol.36, No.2, 1999, pp. 484-486.*

[11] D.W. Bechert, R. Meyer, W. Hage, Drag Reduction of Airfoils with Miniflaps. Can We Learn From Dragonflies?, *AIAA-2000-2315, Denver, CO, 2000.*

[12] M. Schatz, B. Guenther, F. Thiele, Computational Modeling of the Unsteady Wake behind Gurney-Flaps, *2nd AIAA Flow Control Conference, AIAA-2417, Portland, Oregon, USA, 2004.*

[13] Garner, H.C., Rogers, E.W.E., Acum, W.E.A., Maskell, E.C., Subsonic Wind Tunnel Wall Corrections, AGARDograph 109, 1966

[14] E. Wassen, B. Gunther, F. Thiele, J. Marañón Di Leo, J.S. Delnero, J. Colman, U. Boldes. A Combined Numerical and Experimental Study of Mini-Flaps at Varying Positions on an Airfoils. 45th AIAA Aerospace Sciences Meeting and Exhibit, Reno, 8-11 January 2007. Actas del Congreso.

Section 3

Aeroacoustics and Wind Energy

6

Aerodynamic and Aeroacoustic Study of a High Rotational Speed Centrifugal Fan

Sofiane Khelladi, Christophe Sarraf, Farid Bakir and Robert Rey
DynFluid Lab., Arts et Métiers ParisTech
France

1. Introduction

In this chapter we propose a complete, numerical and experimental, analysis of aerodynamics and aeroacoustics of a high rotational speed centrifugal fan. The proposed approach can be extended to any subsonic turbomachine. The main objective of this chapter is to present the state of the art of conducting numerical simulations to predict the aerodynamic and the resulted acoustics of subsonic fans. Different approaches will be discussed and commented.

In our case, we will focus our study on a high rotational speed centrifugal fan, see figure (1). The numerical simulation of aerodynamics is performed using a URANS approach. The numerical results are validated experimentally before using them to supply the aeroacoustic model using a hybrid approach. For the aeroacoustic analysis, first, we will present and use the aeroacoustic analogy based on Ffowcs Williams and Hawkings (FW&H) formalism. And then we will present another advanced approach based on Linearised Euler Equations in which we will take into account reflections as well as rotating sources given by CFD.

2. Aerodynamic study

A shrouded centrifugal fan with high rotational speed and compact dimensions is studied in this paper. A review made by Krain (2005) clarified the state-of-the-arts of the potential development in the field of this kind of centrifugal turbomachinery. The review points out that many machines of moderate efficiency are still in operation but could certainly be improved by making use of recent knowledge in design and computation techniques. Shrouded impellers usually used in high-specific-speed-type centrifugal fans are linked downstream to a vaned diffuser, cross-over bend and return channel. In this type of machine there exists an unavoidable open cavity between the shrouded impeller and the outer casing. The cavity connects the oulet to the inlet of the impeller throught an axial gap and have , then, a significant influence on the performance of the machine.

Concerning the effects of impeller-diffuser interaction on the flow fields and the performance, several studies of this kind of turbomachinery are repported in litterature. It is shown that there always exist flow separations near the diffuser hub which extend further downstream, see Meakhail (2005) works. The flow is strongly three-dimensional with secondary flows on the hub and the shroud of the de-swirl vanes and significant separation occur downstream the cross-over bend, see Khelladi et al. (2005a;b). Consequently the flow is not really axisymmetric and the axial length of the cross-over bend must be increased of nearly 50% to compensate boundary layer blockage. However, there are significant disagreements between computed and measured flow characteristics across the bend. For small sized compressor impellers, the leakage flow becomes major source of performance and efficiency drop as shown by Eum & Kang (2002). Thus, the way to control the leakage flow tends to become of major interest according to Ishida & Surana (2005).

The advanced CFD tools allow now the full optimization of centrifugal fan design, see Kim & Seo (2004); Schleer & Hong (2004); Zangeneh & Schleer (2004) works. For example, a three-dimensional unsteady viscous flow solution was numerically obtained by Khelladi et al. (2005a) by means of powerful computational facilities and robust software, although with very consumptive calculation time. Despite the knowledge accumulated over the past few decades on the design of turbomachinery, the accurate prediction of this kind of flow fields by numerical means is still difficult. This is partly due to the limits of the turbulent models and to the complexity of the flow through such a machine where occur separations and possibly non-axisymmetry having an effect on the performance of the components downstream, see Moon et al. (2003); Seo et al. (2003). There are still significant differences between computations and experimental results, mainly in the prediction of the separation after the bend and its extent.

As far as it concern recent design improvement, a criterion to define cross-flow fan design parameters is reported by Lazzaretto (2003) taking into account the most significant geometric variables affecting performance and efficiency. Other indications are found to design fans according to maximum total pressure criterion, total efficiency, and flow rate, see Lazzaretto et al. (2003); Toffolo (2004). The energy loss process was studied by Toffolo (2004) who compares theory and experiment results concerning performance and efficiency of centrifugal fans. Anyway, the mechanisms and the range of loss generation of centrifugal fans are not documented enough yet.

In this paper, a motor-fan of the kind used a vacuum cleaners was adopted for the study. The centrifugal fan features is characterised by its high specific speed, compact size with a throttling cavity, of interest, between the shrouded impeller and the outer casing. The major flow features in each component have been numerically investigated with a particular attention paid on the cavity. A second series of numerical simulations has been done with a model without cavity to estimate by comparison the effect of the latest. Further more, a basic volumetric loss model was coupled to the numerical model to obtain a cost effective solution. The results are compared with full numerical and experimental results. Finally, a new empirical constant accounting for the effect of the cavity is presented in the last section.

2.1 Numerical approach and experimental setup

2.1.1 Physical model

The motor-fan studied was used to provide large depression capability as required by a vacuum cleaner. The flow downstream is used to cool the motor but due to its complex internal geometry, it is difficult to simulate the flow field combined with the real motor components. Therefore, some assumptions are necessary. Apart from the investigation of flow around the motor structure, the flow through the fan itself is the main concern of research. As shown in figure (1), the studied centrifugal fan consists on the following components: centrifugal impeller with shrouded blades, diffuser and return channel. A diffuser with 17 vanes and a return channel with 8 irregular guide vanes de-swirls the flow. An upstream and downstream ducts are added to the numerical model. In order to better understand the effect of the cavity on the overall performance and the flow structure, a computational model without cavity was configured (Case I), while for the real prototype at practical operating condition, there always exists a clearance between the rotating shrouded impeller and the stationary outer casing (Case II). Geometrical and operating parameters of the centrifugal fan are listed in Tables 1 and 2.

Head, H (m); ΔP(Pa)	1300; 0.159 10^5
Flow rate, Qv (m^3/s)	35×10^{-3}
Rotational speed, N (rpm)	34000
Specific speed $N_s = N\sqrt{Q_v}/H^{3/4}$	29

Table 1. Aerodynamic characteristics at operating point

Description	Impeller	Diffuser	Return channel
Radius of blade inlet (mm)	18	52.7	60
Span of the blade at the entry (mm)	13	6.48	11
Inlet blade angle (°)	64	85	74
Inclination Angle of the blade inlet (°)	85.8	0	0
Radius of blade exit (mm)	52	66.1	33
Span of the blade at the exit (mm)	5.4	8.43	12
Angle of blade exit (°)	64	71.6	15
Inclination angle of the blade exit (°)	0	0	0
Blade number	9	17	8
Blade thickness (mm)	0.8	0.9	1.6

Table 2. Basic geometrical specifications of the centrifugal fan

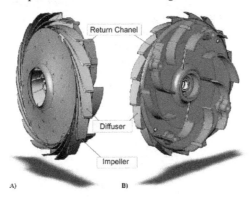

Fig. 1. Description of centrifugal fan system, A) front view, B) back view

2.2 Numerical model

The numerical simulations have been carried out with a code that is based on the finite volume method using FLUENT (1998) to solve the full 3D Reynolds Average Navier-Stokes equations. A centered SIMPLE algorithm is used for the pressure-velocity coupling and a second-order upwind scheme is used for the convection and diffusion terms. The unsteady equations are solved using an implicit second-order upwind scheme. The velocity is specified at the inlet surface for the inlet duct volume and the static pressure is imposed at the exit surface for the outlet duct. Nevertheless, the guide vanes in the return channel are featured as circumferential asymmetry as shown in figure 1. Obviously, to control sound and vibration levels, the number of impeller blades and the number of diffuser vanes have no common divisor. Therefore, for an unsteady solution, the computational domain must cover all the fluid domain. In order to better approximate the near-wall viscous characteristics, the $k - \omega$ SST (Shear Stress Transport) model was adopted. The model treatment close to the wall combines a correction for high and low Reynolds number to predict separation on smooth surfaces, see Menter (1993) paper. Typically, this model gives a realistic estimation of the generation of the turbulent kinetic energy at the stagnation points. The SST model performance has been studied in a large number of cases. The model was rated the most accurate for aerodynamic applications in the NASA Technical Memorandum written by Bardina et al. (1997). As the geometry of the fan is complex, a hybrid mesh is used, composed of tetrahedral elements for the impeller, the diffuser and return channel volumes, and composed of hexahedral elements for upstream and downstream volumes of fluid. A previous study of Khelladi et al. (2005a) shows that a grid of 4.4×10^6 meshes is considered to be sufficiently reliable to make the numerical modeling results independent of the mesh size.

2.3 Measurement methodology

Measurements were carried out on a test bench (see figure 2) equipped with an airtight box (0.6 x 0.6 x 0.6 m), placed upstream of the centrifugal fan. The flow rate from 12 to 60 l/s is controled by changing the diameter of a diaphragm orifice. The unsteady aerodynamic pressure was measured at various positions within the impeller-diffuser unit. Kulite type dynamic sensors, with a diameter of 1.6 mm and a band-width of 125 kHz were used. They are placed at the impeller inlet (figure 2: A), the impeller-diffuser interface (figure 2: B), the diffuser centreline (figure 16:) and the return channel outlet (figure 2: C). This allows the measurement of a static pressure up to 140 mbar and a fluctuating component up to 194 dBA. These aerodynamic data are transmitted to a digital oscilloscope (Gould Nicolet: Sigma 90) with 8 simultaneous channels whose band-width is 25 MHz and has a resolution of 12 bits. Figure 3 presents a schematic of the measuring equipment.

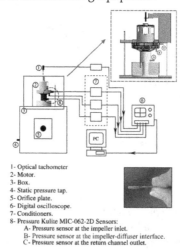

1- Optical tachometer
2- Motor.
3- Box.
4- Static pressure tap.
5- Orifice plate.
6- Digital oscilloscope.
7- Conditioners.
8- Pressure Kulite MIC-062-2D Sensors:
    A- Pressure sensor at the impeller inlet.
    B- Pressure sensor at the impeller-diffuser interface.
    C- Pressure sensor at the return channel outlet.

Fig. 2. Diagram of the test bench

2.3.1 Overall performances and flow field description

Experimental measurements were conducted in order to get the overall and local air-flow characteristics. Two numerical simulations were carried out, the first one does not take into account the cavity between the impeller shroud and the casing, while the second one does. The simulation results relating to the pressure fluctuations in various parts of the fan are compared with the experimental results. For all pressure measurements the uncertainty is ±2.5 mbar.

Figure 3 represents the evolution of the pressure at the inlet of the impeller and the outlet of the return channel according to the flow rate mesured with an uncertainty of ±0.4 l/s. At the outlet of the return channel measurements and numerical results are in good agreement. A phenomenon of blocking produce the increase in static pressure with the flow rate. At the impeller inlet, the calculated pressure curve corresponding to a case without cavity between the impeller and the casing, predicts obviously more depression compared to tests and case with cavity, particularly with flow rates from 21 to 40 l/s. The pressure rise difference between cases with and without gap is due to the presence of strong bakflow vortex at the impeller entrance which leads to the reduction in total and static pressure rise (pressure difference between the inlet and the outlet), see Khelladi et al. (2005a). The effect of vortices decreases at partial flows. The pressure curve calculated taking into account the cavity is in good agreement with the experimental results particularly with flow rates greater or in the vincinity of the nominal flow-rate of 35 l/s. When the fan works at partial flow rates the solution is more difficult to achieve accurately, specially because of the presence of separated flows.

At the nominal flow-rate, the difference of static pressure between experimental result, the case without cavity (Case I) and with cavity (Case II) can be respectively assessed to be of 40% and 3% (experimental results are used as reference). Therefore, it can be considered that the numerical model with the cavity is in adequate accuracy with the measurement and that the effect of the cavity on the overall performance can not be ignored.

The modeling of the cavity, specially the meshing procedure is time and resources consuming. Therefore, when the computation aim the determination of global performance of the fan, volumetric losses due to the cavity and more precisely due to the clearance can be separately modeled using the results of the numerical case without cavity.

The first step is to evaluate the pressure p_f prevailing at the ring seals in view to obtain the volumetric losses and the mass flow rate shift due to the backflow through the cavity. The model is based on simple physical considerations traducing radial balance of the force acting on fluid particles in the volume between the rotating shroud and the fix external casing. For simplification, each part is modeled by a disc. Then, let consider the elementary volume of a ring of fluid of internal radius r, external radius(r+dr) and of lateral length b. This volume is submitted to pressure and centrifugal forces respectively $2\pi \cdot r \cdot b \cdot p$ for inward force, $2\pi \cdot b \cdot (p + dp)$ for outward ($2\pi \cdot dr \cdot b(p + dp)$ is neglectible) and $2\pi \cdot r \cdot dr \cdot b \cdot \rho \cdot \omega^2 \cdot r$ for the centrifugal force. Equilibrium involve force balance to be zero in radial direction. The equilibrium equation is given by,

$$2\pi \cdot r \cdot p \cdot b + \rho \cdot 2\pi \cdot r \cdot dr \cdot \omega^2 \cdot r = (p + dp)2\pi \cdot r \cdot b \qquad (1)$$

and after simplification,

$$dp = \rho\omega^2 \cdot r \cdot dr \qquad (2)$$

The integration of the relation from the exit of the impeller to the clearance located respectively at radius R_2 and R_f provides the expression:

$$p_f = p_2 - \frac{\rho}{8}\omega^2 \left(R_2^2 - R_f^2\right) \tag{3}$$

where p_2 is the static pressure that prevails at this location on the theoretical machine. p_f will be used to estimate the head loss due to the leakage: $\Delta p_f = p_f - p_1$. The velocity of the flow between the static and rotating disc vary from zero to ωR_i. The effective velocity of the flow in the clearance is modeled by $k\omega R_i$. The conventional assumption is made that the flow between the static and the rotating surface has almost the characteristics of a Couette flow and $k \approx 0.5$. This proposal has been verified by a simple CFD calculation in an equivalent machine.

The leakage rate qv_f through the gap is calculated by,

$$qv_f = C_f S_j = C_f 2\pi R_f j \tag{4}$$

where C_f is the actual speed of the fluid in the gap.

The velocity C_f is calculated from the expression (5). The first term on the right end side is the loss due to the contraction and expansion through the gap. Factor 1.5 is empirical. The second term on the right is the loss by friction on the length of the ring where j is the hydraulic diameter. L is the axial length of the clearance and λ the friction loss coefficient for a smooth turbulent flow.

$$\frac{\Delta p_f}{\rho} = 1.5\frac{C_f^2}{2} + \lambda\frac{L}{2j}\frac{C_f^2}{2} \tag{5}$$

with,

$$\lambda = \frac{0.316}{\Re e^{1/4}} \quad \text{and,} \quad \Re e = \frac{C_f 2j}{\nu} \tag{6}$$

The quantity Δp_f is subtracted from the charge of the fan computed without clearance and the flow rate is shifted of qv_f. The final results are confronted figure (3) where it can be seen that the results are in good agreement with experiment on a wider range that the full simulation.

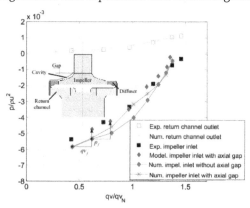

Fig. 3. Comparison of overall performance between calculation and measurement

Figure 4 represents the difference of pressure between experimental measurement, the results of full computation and the results of computation without cavity modified with the model.

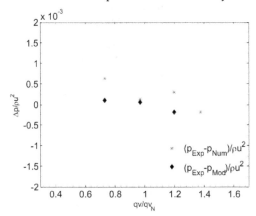

Fig. 4. Pressure difference between measurement and predictions

Figure 5 represents the pressure fluctuation versus time, at point p, as obtained by calculations and tests at the operating point. The uncertainty in time measurement was $\pm 2.5 \times 10^{-6}$ s . Notice that the measurement point is located at the midline of two diffuser blades on the impeller-diffuser interface, so that the blade to blade interaction between the impeller and the diffuser and the separation at the blades leading edge of the diffuser do not disturb the measurement.

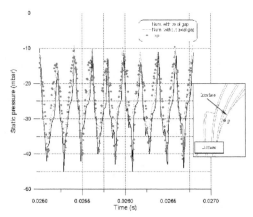

Fig. 5. Variation of the pressure at a point on the impeller-diffuser interface versus time - flow rate = 35 l/s.

On the figure, one distinguishes the impeller blade passage as characterized by pressure peaks. The pressure decreases quickly just after the blade passage and increases gradually with the approach of the following blade. The peak to peak pressure gradient due to the blade impeller passing is very important (30 mbar) compared to (it will be show later) the pressure rise in the entire diffuser. A significant difference between theoretical results with and without cavity is noted at low pressure peaks. The comparison between tests and numerical simulations validates this result. The calculation result taking into account the cavity is closer to the tests than that without. The general shape of the two calculation results are quite similar, the same curvatures are observed at the same time positions. So the axial cavity only affects the peak to peak pressure gradient.

Figure 16 represents the variation of the static pressure along the centerline of the diffuser as measured by flush mounted sensors that do not disturb the flow. The uncertainty in the curvilinear coordinate distance (s) is ±0.1 mm.

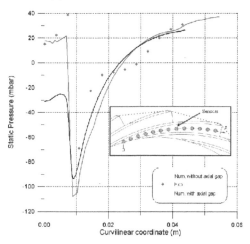

Fig. 6. Evolution of the static pressure in the curvilinear direction along the axial clearance of the diffuser - flow rate = 35 l/s.

The first three points (curvilinear coordinate = 0 to 0.008 m) of measurement in figure 6 are located in a recirculation zone, according to Khelladi et al. (2005a) and the difference between pressure distributions could be seen as a question of static pressure measurment. However, the results of the simulation where the gap is taken into account, is to bee seen mainly in a shift of the pressure rise characteristic. The internal mass flow through the impeller is higher than the mass flow measured at the inlet of the stage due to the leakage flow. This means a given pressure rise across the impeller is attributed to a lower mass flow (internal mass flow trough the impeller minus leakage mass flow). Just after this zone (curvilinear coordinate = 0.008 to 0.01 m) the pressure falls quickly at the leading edge of the diffuser to increase until reaching its maximum value on the outlet side of the diffuser (curvilinear coordinate = 0.01 to 0.045 m). The diffuser transforms a part of the kinetic energy to pressure energy, it increases the pressure from -100 to 30 mbar.

A blade-to-blade section was selected. This sectional view of the static pressure at the section of blade-to-blade at middle radial span (5 mm away from the hub)is shown in figure 7A) and figure 7B) for the Case I and II, respectively. The static pressure difference between the impeller inlet and diffuser outlet can be roughly assessed. The difference for Case I is about twice of the Case II. This implies that the existence of the leakage flow reduces the capability of static pressure rise. The inflow condition to the downstream component(i.e., the diffuser) depends on the flow field at impeller exit. Referring to both cases, it can be found that the flow around one of the diffuser vanes is characterized by the fact that the incidence angle is too positive introducing a strong stagnation point at the suction side near the leading edge, in turn, it produce an acceleration regime in the opposite side. Naturally, this kind of flow increases both the static loading of the diffuser and the force fluctuation due to impeller/diffuser interaction. Not only the through flow capability but also the overall performance will be degraded due to the flow blockage existing in the diffuser.

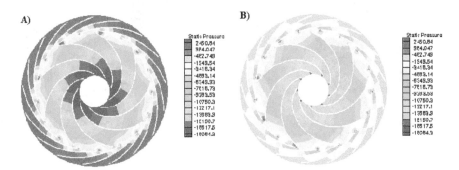

Fig. 7. Comparison of contour of static pressure at a blade-to-blade section of mid radial span for the case, A) without tip clearance and B) with tip clearance

The comparison of the static pressure at the section plotted across the rotating axis for the case with/without cavity is presented in figure 8, figure 8A) and figure 8B) are for Case I and Case II, respectively. The static pressure through the fan inlet and outlet can be roughly figured out for both cases, for Case I it is about 1.3 times bigger than that of Case II. The static pressure is gradually increased from impeller inlet to outlet, inducing a returned flow inside the cavity from the outlet side of impeller back to the inlet and generating a re-circulation flow pattern, see Khelladi et al. (2005a). As can be seen in figure 8B), the static pressure inside the cavity deacrease slowly from the impeller outlet to the gap. At the gap the static pressure of cavity is bigger than that of impeller, which produces an unavoidable source of flow re-circulation distortion ahead of the impeller as well as loss generation.

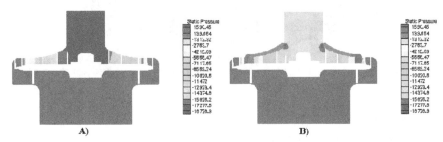

Fig. 8. Comparison of contour of static pressure at a cross-rotating-axis section for the case, A) without tip clearance and B) with tip clearance

The comparison of the total pressure at the cross-rotating-axis section is given in figure 9. The maximum total pressure rise for Case I is roughly 2 times the one of Case II. For Case I, the total pressure is uniformly increased through impeller and diffuser but high distortions are found at the cross-over bend and return channel. For Case II, it exists a non-uniformity at impeller inlet, cross-over bend and return channel. The almost constant total pressure inside the cavity is found to generate a stagnation point near the impeller inlet. As mentioned-above, the cross-over bend and return channel are two dominant components with high loss generation.

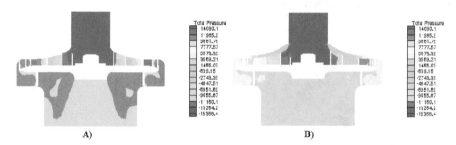

A)   B)

Fig. 9. Comparison of contour of total pressure at a cross-rotating-axis section for the case, A) without tip clearance and B) with tip clearance

Figure 10 shows the solution of entropy for both cases. By this one can be quantitatively understood the loss distribution. For Case I, as shown in figure 10A), very high entropy gradients occur through impeller exit, cross-over bend and return channel. This can also be found in figure 10B) for Case II, typically, a big jump appears at the impeller inlet close to the gap where a strong stagnation point is produced. Due to the high rotating speed, the absolute velocity at impeller exit is so high that a strong swirling flow occurs at the cross-over bend. Due to the narrow frontal area at the cross-over bend and return channel, the strong flow blockage and vortex is accumulated.

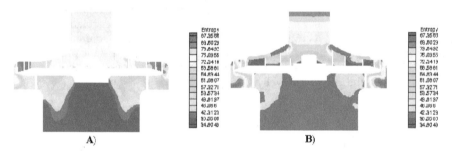

A)   B)

Fig. 10. Comparison of contour of entropy at a cross-rotating-axis section for the case, A) without tip clearance and B) with tip clearance

2.3.2 Loss generation

In order to easily show (quantify) the loss generation for each flow component, a half view of the flow components into the centrifugal fan is schematically plotted as shown in figure 11. The impeller inlet is indicated by A in the figure and the diffuser by B. The cross-over bend is indicated by C. D region is the return channel and the downstream duct is indicated by E.

The vorticity amplitude for each component is normalized by that of the impeller outlet. As shown in figure 12 (Case I), the maximum vorticity amplitude occurs at the impeller due to its high rotating speed. A, B and C regions have equivalent ones with about half of the one of the impeller implying a strong swirling flow and is probably the cause of flow blockages and total pressure drop.

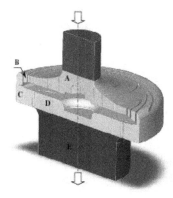

Fig. 11. Schematic of the calculation domain with the flow components of the centrifugal fan (half view)

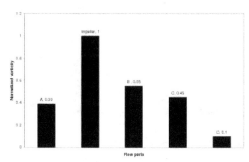

Fig. 12. Distribution of normalized vorticity amplitude along the centrifugal fan (Case I)

The static pressure rise coefficient for the stationary compartment can be written as

$$C_p = \frac{P_{s2} - P_{s1}}{P_{t1} - P_{s1}} \qquad (7)$$

Where subscripts 's' denotes static pressure, 't' the total pressure, '2' the exit section of the component, and '1' the inlet section of the component. The static pressure recovery coefficient along the stationary parts of Case I is shown in figure 13. In this case, the maximum pressure recovery, around 0.53, occurs at the cross-over bend. Both A and D regions have the same pressure recovery, around 0.15.

The loss coefficient for the stationary parts is given by

$$\omega = \frac{P_{t2} - P_{t1}}{P_{t1} - P_{s1}} \qquad (8)$$

In figure (14) one can see that the main part of losses occurs in the return channel and the cross over bend. This is due to the flow separation resulting from the strong meridional curvature and by the 90 degree change of direction of the flow. The comparison between cases I and II is listed in Table 3. We can observe that for Case I the maximum loss occurs at the cross-over

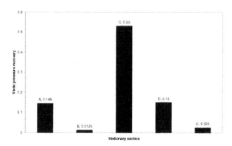

Fig. 13. Static pressure recovery for stationary sections (Case I)

bend while it moves downstream to the return channel for Case II. This is due to the severe flow blockage occurring in Case II.

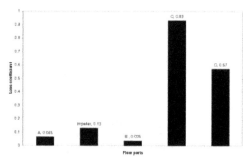

Fig. 14. Distribution of total pressure loss along the centrifugal fan (Case I)

Section	Case I	Case II
A	0.065	0.50
Impeller	0.13	0.14
B	0.035	0.79
C	0.93	0.79
D	0.57	0.97

Table 3. Comparison of loss distribution into the centrifugal fan components

The isentropic efficiency of the impeller is defined as

$$\eta = \frac{\left(\frac{P_{t2}}{P_{t1}}\right)^{\frac{\gamma-1}{\gamma}} - 1}{\frac{T_{t2}}{T_{t1}} - 1} \tag{9}$$

Considering the viscous effect to the total pressure rise, the total pressure coefficient can be expressed as

$$\pi = \tau^{\frac{\gamma-1}{\gamma}} \exp\left(\frac{-\Delta s}{R}\right) \tag{10}$$

where $\tau = 1 + \frac{U_2 C_{\theta 2}}{C_p T_{t1}}$, U_2 is the rotating velocity at the impeller exit, $C_{\theta 2}$ is the circumferential component of the absolute velocity at the impeller exit, T_{t1} is the total temperature at the impeller inlet. Δs is the entropy increment and R is the gas constant. The calculated overall performances for both cases are listed in Table 4. It is found that the total efficiency has 1% drop when the cavity is taken into account, meanwhile, both the total-pressure ratio and temperature ratio drop too.

	Case I	Case II
η (total-total efficiency)	0.87	0.86
π (total-total pressure ratio)	1.15	1.12
τ (total temperature ratio)	1.065	1.057

Table 4. Comparison of overall performance between Case I and Case II

2.4 Concluding remarks

In this section, a rotational speed centrifugal fan with compact dimensions was studied numerically and experimentally. The computational models with/without cavity (between the shrouded impeller and the fixed outer casing) were set up for unsteady solutions. A volumetric loss model was coupled to the case without cavity and compared to the later configurations. The effects of the cavity on the calculated flow field and the loss generation were qualitatively analyzed. The comparison of fan performances between the prediction and measurement showed a favorable tendency. The result obtained by the volumetric loss model, which is easy to setup, coupled to the configuration without cavity gives as good result as the computational model with cavity. This result can be very interesting knowing that the model with cavity is more consumptive in grid size and computational effort than the case without cavity. In terms of the flow restriction obtained for the centrifugal fan, a new empirical constant accounting for the effects of a cavity on the flow restriction of the centrifugal fan system was approximated. In presence of the cavity, due to the difference of static pressure, a leakage flow due the clearance appears and is injected in the main stream at the impeller inlet. Subsequently, a stagnation point appears and changes the incidence angle and then the flow performances as well.

3. Aeroacoustic study

3.1 Computational aeroacoustics for compelex geometries – application to subsonic turbomachines

The sound generation by a flow and its propagation are a matter of aerodynamics. Indeed, the conservation equations of mass and momentum govern both the flow dynamics and the resulting acoustic phenomena. However, the features of the aerodynamic flow and the sound are different. The first is convective and/or diffusive and the second is propagative with very low attenuation due to viscosity. On the other hand, aeroacoustic problems present a wider range of wave-lengths than those of aerodynamic ones.

Aeroacoustic noise optimization is the main topic of many widespread research studies of industrial interest, see Khelladi et al. (2008); Maaloum et al. (2004) works. In fact, the noise level emitted by a device could determine the success or failure of a new prototype. On the other hand turbomachines are widely found in industrial applications. In these devices the level of sound generated is a very important parameter of design.

The prediction of aerodynamic noise benefits from recent developments in numerical methods and computer science. However, despite the knowledge accumulated over the past few decades on the mechanisms of noise generation on complex systems as for example air delivery systems, the prediction of such a flow field and the resulting acoustic pressure, by numerical methods is still difficult. This is due to our inability to model the turbulent viscous flow with enough accuracy on complex geometries and to the complicated nature of flow through turbomachines. Until now, there is still no consensus about the aeroacoustic approach to adopt, and actually, it depends on the application. In the following we present a succinct description of the most commonly used approaches.

Previously to our exposition, we recall the concepts of far and near-fields. The concept of far-field, relative to the effects of the flow compressibility, concerns the propagation of acoustic waves produced by a pressure change in the propagation medium. The occurred disturbance propagates gradually by molecular excitement to the observer far from the source. Unlike the far-field, near-field includes the sound due to the fluid compressibility and another component called the aerodynamic disturbance field or pseudo-sound. It consists of all pressure fluctuations governed primarily by the incompressibility directly related to the flow. These fluctuations are local and are not propagative.

In Computational Aeroacoustics (CAA), two computational approaches are possible:

Direct approach

This approach consists of adjusting the aerodynamic numerical modeling to the acoustics requirements. In other words, it is needed to use numerical schemes adapted to the acoustic propagation, providing low-dissipation and low-dispersion. However, the complexity of implementing these schemes and the far field constraint, where the grid must extend over very large distances, greatly increases the computational costs and makes using this approach very difficult for complex geometries.

Hybrid approach

This approach can be divided into two types of modelling.

- The first one is to use the direct approach near disturbances in which the acoustic waves are propagated over a short distance. They are then propagated using an adapted propagation operator, as Kirchhoff's equation, according to Farassat & Myers (1988), for example, to the far field. For adapted wave operator we mean a wave equation or other conservation equation system that permits an acoustic wave to propagate from a given acoustic source. However, the simulation of the flow field requires DNS or LES, and the treatment of boundary conditions must be done with utmost care to ensure an accurate transition between near and far fields.
- The second consists of separating aerodynamics and acoustics computations. Thus, acoustic sources are given by aerodynamic calculation and propagated using wave equation (Ffowcs Williams & Hawkings (1969), Kirchhoff,...), linearized Euler equations or other approaches like LPCE presented by Moon & Seo (2006),.... The constraints and the computation time is considerably reduced compared to DNS.

Aeroacoustics of complex geometries

Aeroacoustics is a science dealing with the sound generated either by the flow itself, as free jet turbulence or by its interaction with a moving or static surface, rigid or deformable, as fan blades, helicopter rotor, compressors or turbines, etc. Thus, in these latter kinds of applications, we need to deal with flow through complex geometries.

The first attempt to formulate a theory about the acoustics of propellers was conducted by Lynam & Webb. (1919). They showed that the rotation of the blades of a propeller causes a periodic modulation of the fluid flow and associated acoustic disturbance. Another approach, initiated by Bryan (1920), is to study the propagation of a source point in uniform motion. This had as main feature the introduction of the concept of delayed time.

Gutin (1948) was the first to establish a theoretical formalism of a steady noise source through linear acoustics. He showed that steady aerodynamic forces correspond to dipole source distribution on the disc of a propeller. This model proves to be incomplete because, in reality, the noise emitted by rotating blades extends rather high frequencies. The sound at high frequencies is a consequence of the unsteadiness of aerodynamic loads.

Advances in the prediction of noise from the airflow, are based on of Lighthill (1952; 1954) investigations. In his analogy, the generated noise is mathematically reduced to the study of wave propagation in a medium at rest, in which the effect of the flow is replaced by a distribution of sources. The pressure is therefore regarded as characterizing a sound field of small amplitude carried by a fluid, whose properties are uniform throughout the area at rest. The major intake of Lighthill is to include nonlinear terms expressing the noise generation by turbulent flow.

Curle (1955) extended the Lighthill's analogy to include solid boundaries by treating them as distributions of surface loads. Subsequently, Ffowcs Williams & Hawkings (1969) (FW&H) have extended this approach by taking into account the motion of solid surfaces in the flow.

Far and near-fields

The concept of far-field, relative to the effects of the flow compressibility, concerns the propagation of acoustic waves produced by a pressure change in the propagation medium. The occurred disturbance propagates gradually by molecular excitement to the observer located far from the source.

Unlike the far-field, near-field includes the sound due to the fluid compressibility and another component called the aerodynamic disturbance field or pseudo-sound. It consists of all pressure fluctuations governed primarily by the incompressibility directly related to the flow. These fluctuations are local and are not propagative.

3.2 Limits of aeroacoustic analogy and alternative approaches

In the common formulation of aeroacoustic analogy solution, the noise is radiated in free and far field. As strong hypotheses: reflections, diffractions, scattering as well as the confinement effects are not taken into account. These hypotheses make very easy the use aeroacoustic analogy for noise prediction of open rotors or free jets for example, but they are also its weaknesses in case of confining. In paper of Khelladi et al. (2008) it was shown that using the FW&H formulation to model the noise generated by a centrifugal fan does not match measurements because of the presence of a casing. Taking into account the sound attenuation of the casing to correct the directivity has not really solved the problem. For this kind of problems, it was therefore concluded that the aeroacoustic analogy does not obtain accurate results.

To take into account confining effects, it is expected that LEE can give satisfactory results. In fact, with LEE one can use the same acoustic sources as FW&H for example, and in addition reflections, diffractions and scattering are naturally taken into account by an adequate choice of boundary conditions.

In fact, figure (15) shows the predicted overall sound level compared to measurements of the centrigugal fan presented above into the aerodynamic section. The difference between the overall sound level and mesurements is presented in figure (16). As shown in this figure, the difference between measurements and calculation is about 10 dB (over a maximum value of 110 dB) at the radial direction around 90° and 270°, the calculation predict more noise at this direction than measurements. In addition to the assumptions stated above, the reason is that in the theoretical model all the surfaces are considered as acoustically transparent, the effect of the acoustic attenuation of the diffuser blades in the radial direction is not taken into account. In the axial direction, the difference do not exceed 5 dB by taking into account the attenuation of the casing. In this direction the fan radiates in free field.

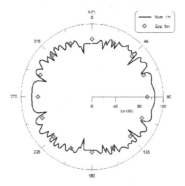

Fig. 15. Numerical and experimental overall noise directivity - $m = 1$, $r_0 = 1$ m, $\varphi = \pi/4$.

Fig. 16. Difference between directivities obtained by the measurements and the calculations ($\Delta L_p = \left| L_{p_{exp}} - L_{p_{num}} \right|$) at $m = 1$, $r_0 = 1$ m and $\varphi = \pi/4$.

To take into account confining effects, it is expected that LEE can give satisfactory results. In fact, with LEE not only one can uses the same acoustic sources as FW&H for example, but also reflexions, diffractions and scattering are naturally taken into account by an adequate choice of boundary conditions.

3.3 Linearised Euler equations

In many aeroacoustic applications we can assume that problems are linear, see Colonius & Lele (2004). In those cases, it is possible to linearize the Euler equations around a (mean) stationary solution $U_0 = (\rho_0, u_0, v_0, p_0)$. Thus, we can write the Linearized Euler Equations written in conservative form are the following:

$$\frac{\partial U}{\partial t} + \frac{\partial F}{\partial x} + \frac{\partial G}{\partial y} + H = S \tag{11}$$

being S a source term and

$$U = \begin{pmatrix} \rho \\ \rho u \\ \rho v \\ p \end{pmatrix} \qquad F = \begin{pmatrix} \rho u_0 + \rho_0 u \\ p + \rho_0 u_0 u \\ \rho_0 u_0 v \\ u_0 p + \gamma p_0 u \end{pmatrix} \qquad G = \begin{pmatrix} \rho v_0 + \rho_0 u \\ \rho_0 v_0 u \\ p + \rho_0 v_0 v \\ v_0 p + \gamma p_0 v \end{pmatrix} \tag{12}$$

$$H = \begin{pmatrix} 0 \\ (\rho_0 u + u_0 \rho)\dfrac{\partial u_0}{\partial x} + (\rho_0 v + v_0 \rho)\dfrac{\partial u_0}{\partial y} \\ (\rho_0 u + u_0 \rho)\dfrac{\partial v_0}{\partial x} + (\rho_0 v + v_0 \rho)\dfrac{\partial v_0}{\partial y} \\ (\gamma - 1)\, p \nabla \cdot v_0 - (\gamma - 1) v \cdot \nabla p_0 \end{pmatrix} \tag{13}$$

where $v = (u, v)$, $v_0 = (u_0, v_0)$, ρ is the density, p the pressure and $\gamma = 1.4$. Subscript $_0$ is referring to mean values. In case of an uniform mean flow, H is null. To solve these equations, we need to compute previously the acoustic sources S, by using LES or DNS. Bogey et al. (2002) presented a methodology to compute the sources.

To solve the above system which is a hyperbolic system, we used FV-MLS method presented and commented by Khelladi et al. (2011). The FV-MLS is a new high order finite volume method for unstructured grids based on moving least squares approximation, it was demonstrated that this method is particularly efficient for this kind of hyperbolic systems. Concerning the numerical developments, we choose to present only the construction of boundary conditions which are one of the most critical points of this kind of problems.

3.4 Boundary conditions

In CAA the treatment of boundary conditions plays a key role according to Colonius et al. (1993), since even small spurious disturbances when the waves leave the domain can distort the acoustic field. In the following we expose our approach to the boundary conditions. For our modeling, the boundary conditions enter in the discretized equations through a proper definition of the numerical flux that can be written as $H(U^+, U^{*-}, n)$, where n points outward from the domain and U^{*-} is the external state variable. Depending on the boundary type, the construction of U^{*-} accounts for, both, the physical boundary conditions that must be enforced and the information leaving the domain.

3.4.1 Reflecting boundary conditions

A perfectly reflecting boundary condition is easily obtained by defining, at each Gauss points on the rigid wall boundaries, an external mirror fictitious state U^{*-}.

The external state is then expressed as

$$U^{*-} = RU^{+} \tag{14}$$

where R is a transition matrix function of n's components, it reads

$$R = \begin{pmatrix} 1 & 0 & 0 & 0 \\ 0 & 1 - 2n_x^2 & -2n_x n_y & 0 \\ 0 & -2n_x n_y & 1 - 2n_y^2 & 0 \\ 0 & 0 & 0 & 1 \end{pmatrix} \tag{15}$$

Using this condition, the mass flux computed by the Riemann solver is zero and the non-permeability condition is satisfied.

3.4.2 Absorbing boundary conditions

Constructing absorbing (non-reflecting) boundary conditions for CAA is pretty delicate because of the high sensitivity of the accuracy to the small spurious wave reflexions at far field boundaries. Approaches based on the characteristics theory are not suited for CAA problems, other approaches, such as Perfectly Matched Layers (PML), presented by Hu (1996), and radial boundary condition of Tam & Webb (1993) are more indicated and widely discussed in the literature for finite differences schemes.

In this work we employ upwinding technique used by Bernacki et al. (2006) with DG to select only outgoing waves at the outer boundaries. Intuitively, it means that whole waves energy is dissipated at boundaries but unfortunately nothing prove that energy is actually dissipated and no spurious wave reflexions persist. To overcome this inconvenient, we join to the above procedure a grid stretching zone, see Nogueira et al. (2010). Grid stretching transfers the energy of the wave into increasingly higher wavenumber modes and the numerical scheme removes this high-frequency content. With this process most of the energy of the wave is dissipated before reaching the boundaries. For a high wavenumber the numerical method introduces more dissipation.

At the grid stretching zone, it is possible to use the MLS method as a filter in unstructured grids. The filtering process is developed by the application of a MLS reconstruction of the variables, i.e:

$$\bar{U}(x) = \sum_{j=1}^{n_{x_l}} U(x) N_j(x) \tag{16}$$

where, U is the reconstructed variable, $\bar{U}$ is the filtered variable and N is the MLS shape function. This reconstruction is performed by using a kernel with shape parameters favoring dissipative behavior that the ones used to the approximation of the variables. The value of these parameters determines the range of frequencies to be filtered.

At the outer boundaries, we propose the following explicit numerical flux,

$$H(U^n, U^{*n}, n) = \frac{1}{2}(\mathbf{F}(U^n) \cdot n + |\mathbf{P}|U^{n-1}) \tag{17}$$

with,
U^{*n} is the fictitious state corresponding to the absorbing side ensuring $\mathbf{P}U^{*n} = |\mathbf{P}|U^{n-1}$, $\mathbf{P}$ is

the Jacobien matrix of system (11) and $|\mathbf{P}| = \mathbf{V}^{-1}|\mathbf{D}|\mathbf{V}$, $\mathbf{D}$ and $\mathbf{V}$ are respectively, eigenvalues diagonal matrix and eigenvectors matrix of $\mathbf{P}$. $|\mathbf{P}|$ is then given by,

$$|\mathbf{P}| = \begin{pmatrix} L_3 & \frac{n_x}{2c_0}(-L_1+L_2) & \frac{n_y}{2c_0}(-L_1+L_2) & \frac{-1}{c_0^2}L_3 + \frac{1}{2c_0}(L_1+L_2) \\ 0 & \frac{n_x^2}{2}(L_1+L_2)+n_y^2L_4 & \frac{n_xn_y}{2}(L_1+L_2-2L_4) & \frac{n_x}{2c_0}(-L_1+L_2) \\ 0 & \frac{n_xn_y}{2}(L_1+L_2-2L_4) & \frac{n_y^2}{2}(L_1+L_2)+n_x^2L_4 & \frac{n_y}{2c_0}(-L_1+L_2) \\ 0 & \frac{n_xc_0}{2}(-L_1+L_2) & \frac{n_yc_0}{2}(-L_1+L_2) & \frac{1}{2}(L_1+L_2) \end{pmatrix} \qquad (18)$$

where,

$$L_1 = |\mathbf{V_0} \cdot \mathbf{n} - c_0|$$
$$L_2 = |\mathbf{V_0} \cdot \mathbf{n} + c_0| \qquad (19)$$
$$L_3 = L_4 = |\mathbf{V_0} \cdot \mathbf{n}|$$

with, $\mathbf{V}_0 = (u_0, v_0)$ and c_0 the speed of sound.

3.5 Application

3.5.1 Validation - convected monopole

This case reproduces the example of Bailly & Juvé (2000). The radiation of a monopole source is computed in a subsonic mean flow, with Mach number $M_x = 0.5$. The source is located at $x_s = y_s = 0$, and is defined as:

$$S_p = \frac{1}{2}\exp\left(-ln(2)\frac{(x-x_s)^2+(y-y_s)^2}{2}\right)\sin(\omega t) \times [1,0,0,1]^T \qquad (20)$$

where the angular frequency is $\omega = 2\pi/30$ and t is the time coordinate. The wave length is $\lambda = 30$ units, and the computational domain is a square with 200 units for each side. The source term is made dimensionless with $[\rho_0 c_0/\Delta x, 0, 0, \rho_0 c_0^3/\Delta x]^T$. With the aim of testing the stability and the behavior of the proposed method for the boundary conditions, an unstructured grid absorbing layer has been added. The absorbing layer is placed from the boundary of the computational domain to $x = \pm 300$ and $y = \pm 300$. In figure 17 (top) it is shown the grid used for the resolution of this problem. To build this grid, 800 equally spaced nodes at the circumference of the computational domain are used and 120 nodes at the outer boundaries circumference.

In addition to the absorbing boundary condition given by equation (17), the shape filter parameters of the absorbing layer are $\kappa_x = \kappa_y = 8$, see Nogueira et al. (2010) paper.

A 5th order mass matrix based FV-MLS solver is used for this example.

Two acoustic waves propagate upstream and downstream of the source, and due to the effect of the mean flow, the apparent wavelength is modified and it is different upstream ($\lambda_1 = (1-M_x)\lambda$) and downstream ($\lambda_2 = (1+M_x)\lambda$) of the source.

In figure 18 pressure isocontours for different non-dimensional times t are shown. The pressure profile along axis $y = 0$ at time $t = 270$ is reproduced in figure 19, and also match the results in Bailly & Juvé (2000).

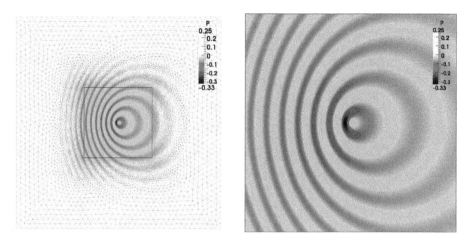

Fig. 17. Convected monopole - M=0.5, 200×200 grid, $t = 270$ at $y = 0$

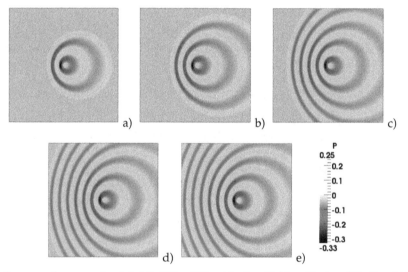

Fig. 18. Convected monopole - M=0.5, 200×200 grid. a) t=60, b) t=90, c) t=150, d) t=210, e) t=270

In order to check the stability of the boundary conditions, we let the computations to continue until 180 periods of the source. This time corresponds to the travel of the wave until the four outer boundaries. Comparing the pressure field with the one corresponding to $t = 270$ (9 source periods), it is observed that there is no change in the solution. The acoustic wave is completely dissipated by teh buffer zone when it lefts the computational domain, see figure (17) (top).

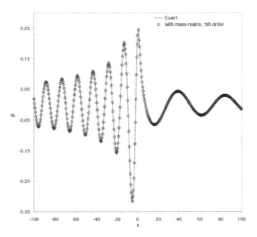

Fig. 19. Convected monopole - M=0.5, 200 × 200

3.5.2 Acoustic waves propagation into a centrifugal fan

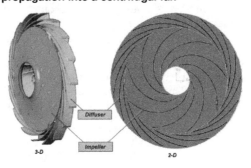

Fig. 20. 3D and 2D centrifugal fan geometric

The centrifugal fan noise is usually dominated by tones produced by the impeller blade passage. The resulted tonal noise corresponds to the blade passage frequency (BPF) and its higher harmonics. This is a consequence of the strong interaction between the impeller and the diffuser blades at their interface.

Shrouded impellers are usually used in high-rotational speed centrifugal fans. The impellers are linked downstream by a vaned diffuser , see figure (20).

A methodology based on a hybrid modeling of the aeroacoustic behavior of a high-rotational speed centrifugal fan is presented in this section. The main objective is to visualize the wave propagation into this machine and demonstrate, then, the power of the proposed methodology. Linearized Euler's equations are used to propagate noise radiated by the rotor/stator interaction. The fluctuating forces at the interaction zone are obtained by an aerodynamic study of the centrifugal fan presented by Khelladi et al. (2005a; 2008). In this section we calculate the acoustic wave propagation of a centrifugal fan with a 9-bladed rotor and a diffuser with 17 blades, as shown in Fig. 13. For the computations we use an unstructured grid, with at least 10 points per wavelength. A detail of the unstructured grid used in this problem is shown in figure (21).

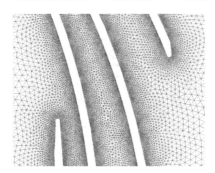

Fig. 21. 2D centrifugal fan mesh

Sources modeling

If we refer to FW&H analogy, one can identify three acoustic sources of three different natures:

- Monopole or thickness source: it is a surface distribution due to the volume displacement of fluid during the motion of surfaces.
- Dipole source or loading source: it is a surface distribution due to the interaction of the flow with the moving bodies
- Quadrupole source: it is a volume distribution due to the flow outside the surfaces.

When the quadrupole source is included, substantially more computational resources are needed for volume integration. However, in many subsonic applications the contribution of the quadrupole source is small. Thus, we have neglected it in this calculation. Moreover, the monopolar source is also neglected at low Mach numbers and small surface thickness.

In our case, the interaction between the impeller and the diffuser blades is the main source of noise radiated by the centrifugal fan, see Khelladi et al. (2008). It is expressed by a pres- sure fluctuation on impeller and diffuser blades. It is, then, of a dipolar nature. This study takes into account only sources located at trailing edge of impeller blades and at the leading edge of blades of diffuser. The rotation of the impeller blades is modeled by rotating sources. Impeller blades are not taken into account in the propagation zone. Thus, we place 17 stationary bipolar acoustic sources located at the leading edge of the blades of diffuser and 9 additional rotating impeller sources located at the trailing edge of each impeller blade.

As for FW&H analogy, the source terms introduced in the LEE are constructed from the momentum equations, and defined by:

$$S_{p_i}(x,y,t) = e^{-\frac{\ln(2)}{2}\left[\left(x-x_{s_i}(x,y,t)\right)^2 + \left(y-y_{s_i}(x,y,t)\right)^2\right]} \times p_i(t) \times \left[0, n_{x_i}, n_{y_i}, 0\right]^T \tag{21}$$

the subscribe i corresponds to blades id, the position of the of the sources are defined by there coordinates $(x_s(x,y,t), y_s(x,y,t))$. For the impeller the sources moves following a circle path, the diffuser sources are static. $p_i(t)$ is the aerodynamic static pressure and (n_x, n_y) are the components of the unit radial vector at sources (x_s, y_s). The exponential term of equation 21 models the punctual nature of the considered sources.

Acoustic pressure history is presented in figure 22. At the beginning of the simulation we can observe clearly the position of sources. But soon we lose track of them because of reflections and interference. Thus, all these effects will be explicitly represented in the far field. Note that they are not represented when other approach is used (FW&H, for example).

Fig. 22. Acoustic pressure history

3.6 Conclusion

In this section, we presented the aeroacoustic analysis of the centrifugal fan studied aerodynamically above. For this kind of machines, the effect of the near-field (casing and confinement) is very important which makes it very difficult to use acoustic analogy techniques. Linearized Euler equations seem to be a good alternative. In this section we presented the first developments and results concerning this approach with complex geometries such as the fan we analyzed. Nevertheless, the effect of rotating blades of the impeller was modeled by rotating sources, only the stator is taken into account. Taking into account the moving parts into the propagation domain is a our challenge for the near future. A technique based on sliding mesh method is under-development, the first results were very promising but some difficulties subsist to maintain the overall space order of accuracy due to the rotor/stator interface stencil reconstructions.

4. References

Bailly, C. & Juvé, D. (2000). Numerical solution of acoustic propagation problems using linearized Euler equations, *AIAA Journal* 38(1): 22–29.

Bardina, J., Huang, P. & Coakley, T. (1997). Turbulence modeling, validation, testing and development, *NASA Technical Memorandum 110446* .

Bernacki, M., Lanteri, S. & Piperno, S. (2006). Time-domain parallel simulation of heterogeneous wave propagation on unstructured grids using explicit, non-diffusive, discontinuous Galerkin methods, *J. Computational Acoustics* 14(1): 57–82.

Bogey, C., Bailly, C. & JuvÃľ, D. (2002). Computation of flow noise using source terms in linearized eulerÂŠs equations., *AIAA Journal* 40(2): 235–243.

Bryan, G. H. (1920). The acoustics of moving sources with application to airscrews., *R. & M. No. 684, British A.R.C.* .

Colonius, T. & Lele, S. (2004). Computational aeroacoustics: progress on nonlinear problems of sound generation., *Prog Aerosp Sci* 40: 345–416.

Colonius, T., Lele, S. & Moin, P. (1993). Boundary conditions for direct computation of aerodynamic sound generation, *AIAA Journal* 31(9): 1574–1582.

Curle, N. (1955). The influence of solid boundaries upon aerodynamic sound, *Proceedings of the Royal Society of London* A231: 505–514.

Eum, H.-J. & Kang, S.-H. (2002). Numerical study on tip clearance effect on performance of a centrifugal compressor, *Proceedings of ASME FEDSM'02, Montreal, Quebec, Canada, July 14-18* .

Farassat, F. & Myers, M. K. (1988). Extension of kirchhoff's formula to radiation from moving surfaces, *J. Sound and Vib.* 123: 451–560.

Ffowcs Williams, J. & Hawkings, D. (1969). Sound generation by turbulence and surfaces in arbitrary motion, *Philosophical Transactions for the Royal Society of London* A264: 321–342.

FLUENT (1998). Fluent, inc.

Gutin, L. (1948). On the sound field of a rotating propeller, *NACA TM1195, (Traduction de "Über das Schallfeld einer rotierenden Luftschraube", Physikalische Zeitschrift der Sowjetunion, Band 9, Heft 1, 1936).* pp. 57–71.

Hu, F. (1996). On absorbing boundary conditions for linearized Euler equations by a perfectly matched layer, *Journal of Computational Physics* 129: 201–219.

Ishida, M. & Surana, T. (2005). Suppression of unstable flow at small flow rates in a centrifugal fan by controlling tip leakage flow and reverse flow, *Transaction of the ASME Journal of Turbomachinery, Vol. 127, Jan., 76-83* .

Khelladi, S., Kouidri, S., Bakir, F. & Rey, R. (2005a). Flow study in the impeller-diffuser interface of a vaned centrifugal fan, *Journal of Fluid Engineering* 127: 495–502.

Khelladi, S., Kouidri, S., Bakir, F. & Rey, R. (2005b). A numerical study on the aeroacoustics of a vaned centrifugal fan, *ASME Heat Transfer/Fluids Engineering Summer Conference, Houston, Texas* FEDSM2005-77134.

Khelladi, S., Kouidri, S., Bakir, F. & Rey, R. (2008). Predicting tonal noise from a high rotational speed centrifugal fan, *Journal of Sound and Vibration* 313, Issues 1-2: 113–133.

Khelladi, S., Nogueira, X., Bakir, F. & Colominas, I. (2011). Toward a higher order unsteady finite volume solver based on reproducing kernel methods, *Computer Methods in Applied Mechanics and Engineering* 200(29-32): 2348 – 2362.
URL: *http://www.sciencedirect.com/science/article/pii/S0045782511001344*

Kim, K. & Seo, S. (2004). Shape optimisation of forward-curved-blade centrifugal fan with navier-stokes analysis, *Journal of Fluid Engineering* 126: 735–742.

Krain, H. (2005). Review of centrifugal compressor's application and developement, *Journal of Turbomachinery* 127: 25–34.

Lazzaretto, A. (2003). A criterion to define cross-flow fan design parameters, *Journal of Fluids Engineering, Vol.125, 680-683* .

Lazzaretto, A., Toffolo, A. & Martegani, A. D. (2003). A systematic experimental approach to cross-flow fan design, *Transaction of the ASME Journal of Fluids Engineering, Vol. 125, July, 684, 693* .

Lighthill, M. J. (1952). On sound generated aerodynamically, i. general theory, *Proceedings of the Royal Society of London* A211: 564–587.

Lighthill, M. J. (1954). On sound generated aerodynamically, ii. turbulence as a source of sound, *Proceedings of the Royal Society of London* A222: 1–32.

Lynam, E. & Webb., H. (1919). The emission of sound by airscrews., *R. & M., No. 624* .

Maaloum, A., Kouidri, S. & Rey, R. (2004). Aeroacoustic performances evaluation of axial fans based on the unsteady pressure field on the blades surface, *Applied Acoustics* 65: 367–384.

Meakhail, T. (2005). A study of impeller-diffuser-volute interaction in a centrifugal fan, *Transaction of the ASME Journal of Turbomachinery, Vol.127, Jan., 84-90* .

Menter, F. (1993). Zonal two equation $k - \omega$ turbulence models for aerodynamic flows, *AIAA Paper* 93-2906.

Moon, Y.-J., Cho, Y. & Nam, H.-S. (2003). Computation of unsteady viscous flow and aeroacoustic noise of cross flow fans, *Computers & Fluids, Vol. 32, 995-1015* .

Moon, Y. & Seo, H. J. (2006). Linearized perturbed compressible equations for low mach number aeroacoustics., *Comput. Phys.* 218(2): 702–719.

Nogueira, X., Cueto-Felgueroso, L., Colominas, I., Khelladi, S., Navarrina, F. & Casteleiro, M. (2010). Resolution of computational aeroacoustics problem on unstructured grids with high-order finite volume scheme, *Journal of Computational and Applied Mathematics* 234 (7): 2089–2097.

Schleer, M. & Hong, S. (2004). Investigation of an inversely designed centrifugal compressor stage - part ii: Experimental investigation, *Journal of Turbomachinery* 126: 82–90.

Seo, S., Kim, K. & Kang, S. (2003). Calculations of three-dimensional viscous flow in a multiblade centrifugal fan by modeling blade forces, *Journal of Power and Energy* 217: 287–297.

Tam, C. & Webb, J. (1993).   Dispersion-relation-preserving finite difference schemes for computational aeroacoustics, *J. Comput. Phys.* 107: 262–281.

Toffolo, A. (2004).   On cross-flow fan theoretical performance and efficiency curves: An energy loss analysis on experimental data, *Transaction of the ASME Journal of Fluids Engineering, Vol. 126, Sept., 743-751* .

Zangeneh, M. & Schleer, M. (2004).   Investigation of an inversely designed centrifugal compressor stage - part i: Design and numerical verification, *Journal of Turbomachinery* 126: 73–81.

7

Wind Turbines Aerodynamics

J. Lassig[1] and J. Colman[2]
[1]*Environmental Fluid Dynamics Laboratory,*
Engineering Faculty, National University of Comahue
[2]*Boundary Layer and Environmental Fluid Dynamics Laboratory,*
Engineering Faculty, National University of La Plata
Argentina

1. Introduction

Since the earlier petroleum crisis (decade of 70s), when begun the interest of the aeronautical industry into wind turbines development until the present, 40 years of research and development become in an important design evolution.

At the present we could say that a modern rotor blade design implies some different aerodynamics criteria than used regarding wings and airplane propellers designs.

This is due that their operating environment and operation mode are quite different than airplanes ones. Such differences could be summarized as follows:

1. Flow characteristics of the media where they function
2. Relative movement of the aerodynamic parts regarding free upwind flow

1.1 Environment

Wind turbines are immersed in the low atmospheric boundary layer, characterized by the inherent turbulent nature of the winds and, also, for the presence of dust, sand and insects, which finally ends gluing to the blades surfaces incrementing their roughness.

For the reasons mentioned above, the rotor blade airfoil is submitted to a time and space variation flow, producing on it different superimposed phenomena promoting flow hysteresis, like dynamic stall.

By other way, the wind which transports dust and other particles in the boundary layer, will change the apparent blade roughness and for that reason the airfoils to be used on blade design, should be almost no-sensitive to such roughness changes.

1.2 Operating mode

The second aspect to be considered is the rotor operating mode, because blade turbines have a relative movement regarding upwind flow, with changes at each blade section due the resultant velocity at the blade will be the vector sum between the upwind flow and the tangential rotation.

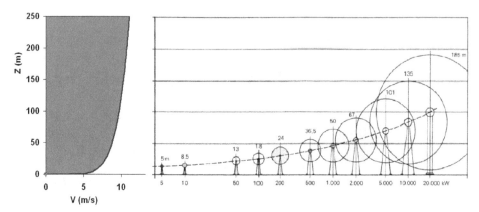

Fig. 1. Typical vertical mean wind velocity distribution in the low atmospheric boundary layer, compared with the rotor size evolution

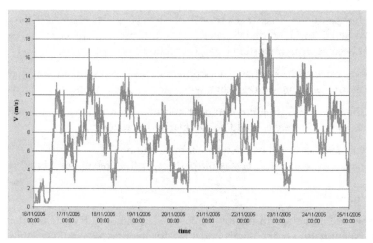

Fig. 2. Wind time variation during a 10 days period, measured in a 30m height tower located at Vavarco town, Neuquen´s Province, Argentina

2. Wind turbines

The energy per unit time transported by the upstream wind, or meteorological power, is defined as:

$$P_m = \frac{1}{2} \cdot \rho \cdot \int_0^\infty f(v) \cdot V^3 dv \qquad (1)$$

where $f(v)$ is the wind distribution function at the zone to be considered.

That´s the indicative wind power of the potentiality of the zone. In order to quantify it in watts, we must multiply such formula by the area to be considered ($A = \pi \cdot R^2$), achieving by this way the expression for the Available Power:

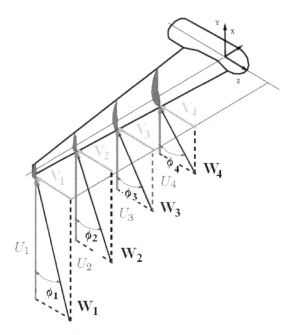

Fig. 3. Resultant flow over rotor blades, being V the mean free upwind velocity, U the tangent velocity, W the resultant and φ the effective pitch angle, measured respect the rotation plane

$$P_d = 1/2 . \rho . V_0^3 . \pi . R^2 \tag{2}$$

In order to extract all that power, by means of the rotor, the wind velocity behind it should be zero and, obviously, that´s impossible. So, we could only extract part of the Available Power being such part the Obtainable Power:

$$P_{obt} = P_d . C_p \tag{3}$$

Where C_p is the rotor´s power coefficient.

2.1 Momentum or Froude´s or Betz´s propeller model

This simply model consists on considerer the rotor as an *actuator disk* integrated by a great number of infinitesimal width needles. The disk rotate when the wind pass trough it. The hypotheses of the model are:

a.   Upstream flow is non-rotational, non-viscous and incompressible
b.   Air, at the disk rotation plane is rotational, viscous, etc, but will not be considered directly on the calculus.
c.   Downstream flow will be non-rotational, non-viscous and incompressible but, due it crossed the disk, the Bernoulli constants will be different upstream and downstream. This constant change is the only factor which "incorporates" part of the true behavior of the flow on crossing the disk plane.

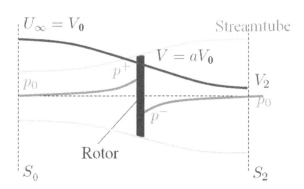

Fig. 4. Expansion stream tube effects, upon the flow through the rotor, producing a downwash velocity reduction (V_2).

The upwind stream tube have a velocity V_0 and, after flow through the rotor plane, part of the kinetic energy is transformed in extracted energy due the rotor, producing a downwind velocity reduction and, so, the stream tube will expand in order to fulfill the continuity equation.

V_0 and V_2 are defined as:

$$V = V_0 (1-a) \tag{4}$$

$$V_2 = V_0 (1-2a) \tag{5}$$

Being a an axial induction factor, which take into account the stream tube expansion. According that, a non-dimensional Power Coefficient C_p is defined:

$$C_p = \frac{AvailablePower}{1/2 \cdot \rho \cdot V_0^3 \cdot A} \tag{6}$$

$Available\ Power$= Force .Velocity = $\Delta P.A.V = (P^+ - P^-).A.V = \frac{1}{2}\rho AV(V_0^2 - V_2^2)$ (7)

Using [4] and [5] in the former equation, we obtain:

$$Available\ Power = 2.\rho.A.a(1-a)^2.V_0^2 \tag{8}$$

Then C_P could be expressed as:

$$C_p = 4 \cdot (1-a)^2 \cdot a \tag{9}$$

Taking in account the blades rotation (angular velocity w_d) and that the air, on trespassing the rotor plane, receive also a rotation movement w_w, slightly different than the blade rotation (see Figure 5), we could relate both rotation velocities as

$$w_w = a' . w_d \tag{10}$$

Being a' an rotational induction factor. According all those factors, the expression for the Power Coefficient will be:

$$C_P = 4 a (1-a)^2 / (1+a') \tag{11}$$

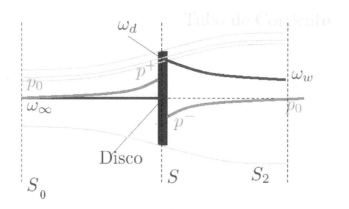

Fig. 5. Stream tube changes, where we could appreciate the rotation effects of the upwind (w$_\infty$), on the rotor plane (w$_d$) and downwind (w$_w$)

2.2 Blade element theory

The aerodynamic forces upon the rotor blades could be expressed as functions of lift and drag coefficients and the angle of attack.

In doing so, we divide the blade in a finite but large number of sections (N), denominated blade elements.

The theory is supported by the following assumptions: there aren´t any aerodynamic interaction between each blade element (which is equivalent as assume 2D flow over the blade); the velocity components over the blade along wingspan aren´t take in account; forces upon each blade element are determined by the aerodynamic airfoil characteristics (2D flow). All of these implies, in fact, that the blade´s aspect ratio will be great than 6.

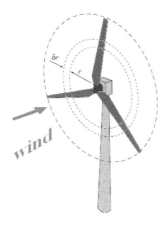

Fig. 6. Blade element of width δr, located at a distance r from the rotor hub. Note the annulus traverse surface due the rotation.

In the blade-element analysis, section lift and drag are normal and parallel to the relative wind, respectively (see Figure 7). It´s consider a turbine blade sliced in N elements of chord C, width dr, with geometric pitch angle ß between rotor´s rotation plane and the chord line for zero lift.

Both, chord and pitch, will vary along blade wingspan. If Ω is the angular velocity and V_∞ the mean upstream wind velocity, the angular component of the resultant velocity at the blade will be:

$$(1 + a') \Omega . r \qquad (12)$$

and the axial component:

$$(1 - a)V_\infty. \qquad (13)$$

So, the relative velocity in the blade plane will be expressed as:

$$W = \sqrt{V_\infty^2(1-a)^2 + \Omega^2 r^2 (1+\acute{a})^2} \qquad (14)$$

This relative velocity forms an angle a respect the rotation plane.

From definitions and showing Figure 7, we could write:

$$\tan\phi = \frac{V_\infty(1-a)}{\Omega r(1-\acute{a})} = \frac{(1-a)}{(1-\acute{a})\lambda_r} \qquad (15)$$

Net force at each blade-element, normal to the rotation plane, could be expressed as

$$\delta F = r(\delta L \cdot \cos\phi + \delta D \cdot \sin\phi) \qquad (16)$$

Being δL and δD the differential lift and drag, respectively, and L and D the total lift and drag.

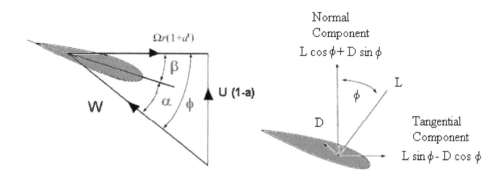

Fig. 7. Velocities and forces acting upon a blade-element

The differential torque will be:

$$\delta Q = r(\delta L \cdot \sin\phi - \delta D \cdot \cos\phi) \tag{17}$$

Such expressions will be employed to calculate the axial and tangential induction factors in the combined theory Blade-Element – Momentum in section 2.4.

2.3 Tip losses account

Due pressure on the blade´s upper surface (or suction side) is, in general, less than on the lower surface (high pressure side), air tends to move from the high pressure side to the lower pressure one (similar to a wing), producing a vortex system at the trailing edge called "trailing edge vortex system" which, combined with the pressure distribution on both sides, generate a lift distribution that tends to zero at the tips. All of that is responsible of the so called "tip losses", in a similar way than the airplane wing. Precisely, such 3D flow pattern is characteristic of a blade rotor or a wing, being the fluid dynamic difference between a wing and an airfoil in which the flow is 2D.

Stall is proper of 3D flow and can´t be determined by the blade-element theory. Nevertheless there are some physical models designed to include the tip losses. The most used of such models was developed by Prandtl (see also Glauert, 1935).

According such model a factor F is introduced in the equations to calculate net force and torque.

Such factor is a function of the number of blades B, effective pitch angle ϕ, blade element coordinate r and blade length, R.

Their mathematical empirical equation is:

$$F = \frac{2}{\pi}\cos^{-1}\left[\exp(-\frac{BR-r}{2r\sin\phi})\right] \tag{18}$$

2.4 Combined theory of momentum with blade element (BEM theory)

The purpose of this theory is to achieve a usable model to evaluate axial (a) and tangential induction factors (a´), from the equations for thrust and torque previously deduced by blade element theory. Now, the actuator disk will be an annulus of width δr and the differential lift and drag forces acting on it will be:

$$\delta L = \frac{1}{2}\rho \cdot W^2 \cdot c_L \cdot b \cdot \delta r \tag{19}$$

$$\delta D = \frac{1}{2}\rho \cdot W^2 \cdot c_D \cdot b \cdot \delta r \tag{20}$$

Replacing the previous equations in the [16] and [17] equations, results:

$$\delta Q = r \cdot \frac{1}{2} \cdot \rho \cdot W^2 \cdot b \cdot (c_L \cdot \sin\phi - c_D \cdot \cos\phi) \tag{21}$$

$$\delta F = \frac{1}{2} \cdot \rho \cdot W^2 \cdot b \cdot (c_L \cdot \cos\phi + c_D \cdot \sin\phi) \tag{22}$$

By other way we could express δF and δQ as:

$$\delta Q = r \cdot \frac{1}{2} \cdot \rho \cdot W^2 \cdot b \cdot c_Q \tag{23}$$

$$\delta F = \frac{1}{2} \cdot \rho \cdot W^2 \cdot b \cdot c_F \tag{24}$$

Equating the equations [8] and [16], for B blades, we obtain

$$2\rho A V_0^2 a(1-a) = B(l \cdot \cos\phi + d \cdot \sin\phi)R = BR\frac{1}{2}W^2 b \cdot C_F \tag{25}$$

Introducing solidity (σ) :

$$\sigma = \frac{B \cdot b \cdot R}{\pi \cdot R^2} = \frac{B \cdot b \cdot R}{A} = \frac{B \cdot b}{\pi \cdot R} \tag{26}$$

Being σ the relation between the area of B blades and the area described by the rotor; b is the blade chord.

So, with the help of the Prandtl correction factor, the induction factors will result:

$$a = \frac{1}{\dfrac{4 \cdot F \cdot \sin^2\phi}{\sigma \cdot c_F} + 1} \tag{27}$$

$$\acute{a} = \frac{1}{\dfrac{4 \cdot F \cdot \sin\phi \cdot \cos\phi}{\sigma \cdot c_Q} - 1} \tag{28}$$

2.5 Modifications to the above classical theory (BEM modifications)

Classic theory, developed in section 2.4, bring good results under operative conditions near specific velocity design, where specific velocity λ is the quotient between blade´s tangential tip velocity and free stream upwind velocity.

Nevertheless, for such specific velocity somewhat far from the design value, for example, for too high or too small λ values, classical theory didn´t give very good results. But, why is it so? Because the theory doesn´t take in account the 3D flow nature, turbulence, stall or losses.

In other words, classical theory work very good in situations where 3D flow is not too pronounced, but this isn´t always the situation and some modifications should be introduced to the theory.

Flow separation at the tip blade, promotes a downstream lowering of the static pressure Also, high static pressure appears on the stagnation zone of the blade. Such pressure differences produce high thrust values on the blades that aren't predicted by the classical BEM theory.

Glauert (1926, 1935) analyzed different induction factors (a) according flow pattern and propeller types. Spera (1994) proposed a rotor thrust taking in account the induction factor, a (see Figure 8).

According flow patterns, it's possible to identify two different turbine states: one, called windmill or slightly charged state, is when turbulence doesn't dominate flow field; the other, with high turbulence or high charged state, and classical BEM theory fails.

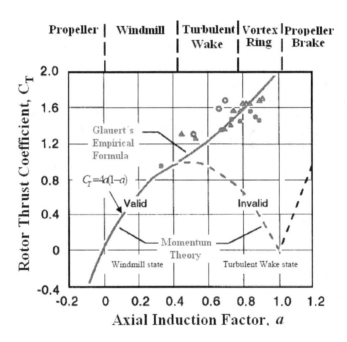

Fig. 8. Experimental and predicted C_T values

Phenomena described above could be included to the BEM theory, by different models: C_T and/or "a" predictions.

Glauert, in 1926, developed a correction for such phenomena employing experimental data from helicopters rotor blades. Such model was thought to correct overall thrust coefficient. Nevertheless, could be used to correct local thrust coefficient by means of the BEM theory.

Basically, Glauert corrections are related with tip losses model. When they grow up, also induced velocities so, hence the turbulence at the wake will increase. According that, the induced velocity calculus must take in account a combination of tip losses and Glauert correction.

Moriarty and Hansen (2005) in their book, explain the empirical Glauert relation modified with the tip losses factor.

The model is

$$C_T = \frac{8}{9} + (4F - \frac{40}{9})a + (\frac{50}{9} - 4F)a^2 \tag{29}$$

$$a = \frac{18F - 20 - 3\sqrt{C_T(50 - 36F) + 12F(3F - 4)}}{30F - 50} \tag{30}$$

2.6 Rotational effects upon aerodynamic coefficients

Himmelskamp (1947) investigated the increments of the maximum lift coefficients for rotating airplane propellers, assuming a radial flow downstream them. He found that such increment in the maximum lift is more pronounced for small radial sections, than the found values in the non rotation state at high angles of attack.

Other researchers have investigated the rotation effects for helicopters. Such investigations assume a local oblique flow at the rotor plane, during forward fly. In the work performed by Harris (1966), rotation effect is assumed by a yaw flow on the rotor plane. He reported that on calculating aerodynamic coefficients by conventional procedures, with a flow component normal to the blade axis, maximum lift also is enhanced with the oblique flow.

Fundamental aspects of the flow configuration in the model, is that the vorticity axis isn´t normal to the local flow direction and the separated flow is transported in the wingspan direction.

For forward flying helicopters, Dwyer & McCroskey (1970) offered a good description of the cross flow effects in the rotor blades plane.

At begin of the power control due aerodynamic stall in wind turbines, were observed that when such control actuated during stall, power tends to exceed the design value.

To describe rotational effects or delayed stall, various models were formulated, consisting the majority of them on add or extend lift coefficient without rotation. There are also few models which introduce corrections to the drag coefficient.

Some rotation effects correction models, are based upon pump air by centrifugal methods, over the separation bubble near trailing edge.

One of the first of such models was described by Sørensen (1986). In his work he showed flow patterns with radial configuration in the separation area around trailing edge.

His experimental results were reproduced, by himself and collaborators, by computational methods on S809 airfoil (Sørensen et all, 2002), which showed clearly centrifugal pumping at the trailing edge.

2.7 Centrifugal pumping mechanism

Centrifugal pumping on the separated air near trailing edge, originate radial flow. The result is that separation bubble size is diminished respect the situation of no-centrifugal loads. Due centrifugal load gradient, along wingspan, air pressure in the bubble is lower and, hence, normal force on the airfoil is enhanced.

At high angles of attack, pressure distribution on the upper surface of the airfoil, produce a suction peak just behind leading edge (leading edge suction force), which diminishes through the trailing edge. The magnitude of such force is proportional to dynamic pressure and, hence, it grows with the square of radial position.

Dynamic pressure gradient along wingspan and negative gradient along chord, conforms a mechanism responsible of the air, at the stall zone, flows to important radial coordinates and could overcome Coriolis forces.

Klimas (1986) describes radial flux by means of Euler equations, including centrifugal effects and Coriolis over separation bubble near the trailing edge.

After that, Eggers y Digumarthi (1992), and other authors, called such mechanism "centrifugal pumping".

Also was reported as "radial pumping" (Sørensen E.A., 2002) or "pumping along wingspan" (Harris, 1966).

Due centrifugal pumping on the separated mass flow around trailing edge, produce additional negative values to the pressure coefficient on the airfoil´s surface.

This extra "negative pressure" gives a negative gradient along chord which is favorable for the boundary layer stability and, in consequence, could shift separation point toward trailing edge. This change in the separation point is hard to model.

Nevertheless, we could assume that separation occurs at a higher angle of attack, being the pressure gradient along chord (included the increment due rotation effects) the same as the state without rotation.

Based upon that relation, it´s possible to elaborate correction models to take in account the increment of the angle of attack (delay).

In relation with such angle of attack change, lift coefficient should be extended in order to have the same slope between curves for totally attached flow (potential one) and totally separated.

Himmelskamp (1947) reported that if the lift stall coefficient, at high angles of attack, is delayed (or shifted), there aren´t observed drag coefficient increment.

Some physical models, which take in account rotation effects, are formulated upon hypothesis of "stall delay", like the ones of Corrigan and Schillings (1994).

2.8 Stall delay model by Corrigan and Schillings

Corrigan and Schillings (1994) developed a model to take in account correction effects upon rotation, expressed in terms of lift coefficients delay for high angles of attack.

The development of this method, begin with boundary layer equations following Banks & Gadd (1963).

Together with the expression for the boundary layer velocity gradient $\partial u / \partial z$, the delay amount was related with angular position of detachment point (θ_S). Precisely, problem formulation in terms of angular detachment point implies dependence between chord/radius relationships similar to other models.

One of the characteristic assumptions of the above cited model is that airfoil without rotation with maximum lift, could have a pressure suction peak at the leading edge, originating an important radial pressure gradient and, so, an important radial flux.

Due simplicity reasons, finally Corrigan and Schillings formulated their model in terms of the trailing edge angle θ_{TE}. For chord values not too big, that could be expressed in terms of the relation chord/radius (c/r).

Stall delay is expressed in terms of a change of the angle of attack, for lift coefficients without rotation:

$$\Delta \alpha = / \alpha_{CL\,max} - \alpha_{CL=0}) \cdot ((\frac{K \cdot \theta_{TE}}{0.136})^n - 1) \qquad (31)$$

Factor K describes the velocity gradient according the universal relation:

$$c/r = 0.1517 / K^{1.084} \qquad (32)$$

For n = 0 (see equation [31]), the above formulas give aerodynamic coefficients without rotation. Corrigan established that for values of n between 0.8 and 1.6, there are a good correlation with the existing experimental data and, a value of 1 match with very good results in many situations. Authors like Tangler & Selig (1997) and Xu & Sankar (2002) used n=1.

The increment of lift coefficient (without rotation) is expressed in terms of the angle of attack increment, as:

$$C_{L.rot} = C_{L.non-rot(\alpha+\Delta\beta)} + (\partial C_L / \partial \alpha)_{pot} \Delta \alpha \qquad (33)$$

Here, $(\partial C_L / \partial \alpha)_{pot}$ is the slope in the linear part of the polar lift coefficient vs. angle of attack. For such, Xu & Shankar used n=0.1.

2.9 Computational solvers

There were developed many CFD solvers for wind turbine blades design and, just for information only, we´ll mention 4 of them developed in USA, in a period of 15 years, presenting the evolution in the solvers content.

3. Aerodynamic airfoils

Airfoil to be selected on the rotor blade design, according our comments at Section 1, should satisfy requirements which are different than those used in the wing design for standard airplanes.

3.1 Reynolds number

Reynolds numbers upon blades are low and vary from the root to the tip. Table 2 shows typical Reynolds numbers corresponding to three turbine power levels, from root and tip. We could observe just only for wind turbines of big power (megawatt type), Reynolds number exceeds 10^6.

CODES Features	NUPROP AeroVironment	PROP93 AEI	PROPID Univ. of Illinois	AeroDyn NREL
Development Date	1986	1993	1997	2002
Airfoil Data Interpolation	no	no	yes	yes
3-D Stall Delay	no	no	yes	yes
Glauert Approximation	yes	yes		yes
Tip Losses	yes	yes	yes	yes
Windspeed Sweep	yes	yes	yes	yes
Pitch Sweep	yes	yes	yes	yes
Shaft Tilt	yes	yes	yes	yes
Yaw Angle	yes	yes	yes	yes
Tower Shadow	yes	no	yes	yes
Dynamic Stall	yes	no	no	yes
Graphics	no	yes	no	yes
Program Language	Fortran	C	Fortran	Fortran
Other	turbulence	hub ext.	Inverse design	The generalized dynamic wake theory

Table 1. Blade turbines CFD solvers time evolution in USA

Power	Re root	Re tip
50 kW	0.2 M	1.3 M
600 kW	0.7 M	1.3 M
2 MW	1.5 M	2.2 M

Table 2. Reynolds numbers for root and tip of wind turbines blades, for different power.

3.2 Low Reynolds number airfoils

First wind turbine developments used NACA airfoils like 4415, 4418, 23012, 23024, etc, which have good aerodynamic characteristics for higher Reynolds.

Boundary layer over a common airfoil, like NACA 4-digits series, under low angles of attack, have usually a favorable pressure gradient from the front stagnation point to approximately the maximum airfoil's width, being the flow in most cases, laminar over such part. After the zone of minimum pressure, the pressure gradient changes its sign becoming unfavorable. Such sign change take place, in general, in a short distance.

Because that, for Reynolds number – based upon airfoil chord and mean free upstream velocity – below 5×10^5, boundary layer separate and again attach conforming a recirculation bubble on the upper surface of the airfoil. Such bubble could be short or long, depending on the perturbations present at the separation point and the local Reynolds number based upon boundary layer momentum thickness and local velocity, exactly before such point. See, for example, Chandrsuda & Bradshaw (1981) and Gad-el-Hak (2000).

Such changes in the flow on the upper surface affects negatively the lift generation. In fact, lift is especially sensible to any perturbation of the flow on the upper surface in higher degree than on the lower surface. When boundary layer begins its detachment, conforming after that the mentioned recirculation bubble, the flow over detachment area becomes a shear layer which, downstream, evolution as a mix layer. Shear layer stability is particular sensible to velocity profile. The whole picture is a recirculation bubble with a shear layer on it, which could or not reattach (Ho & Huerre, 1984).

The main cause of bubble formation is the almost abrupt change in the pressure gradient on the upper surface, from favorable to unfavorable in a small extension. To minimize those adverse effects, appeared the low Reynolds number airfoils (Zaman y Hussain, 1981), (Eppler y Somers, 1980), (McGhee y Beasley, 1973), (Eppler, 1990), (Carroll, Broeren, Giguere, Gopalarathnam, Lyon and Selig, 1990-2000), being their main characteristic the gradual change of pressure gradient sign from negative to positive in a larger extension than the standard airfoils, for a $5x10^4 \leq Re \leq 5x10^5$ range. So, a better lift distribution was achieved.

Nevertheless, all mentioned airfoils designs were tested in laminar flow wind tunnels, but the real condition of the flow where are wind turbines operate is in general turbulent, that's, the flow corresponding to the low atmospheric turbulent boundary layer. Some researchers, like Delnero et al (2007), have performed experiments in boundary layer wind tunnels, with low Reynolds airfoils. They studied the flow field at the neighbor of the airfoil and, in particular, how is the bubble formation and evolution, under turbulent free upstream.

Some wind turbine builders, apart from low Reynolds airfoils, tested laminar airfoils like NASA LS(1)_0413 MOD but, due their high sensibility to roughness their required repeatedly clean due insects upon blade surface.

For such reason, researchers realized that one of the most important characteristic of the airfoils to be used on rotor blade design, should be those with less sensibility to roughness changes during their operation, by dust and/or insects on the leading edge area.

Research performed in USA and Europe leave to design of new airfoils families which are less sensible to roughness changes.

For example, the National Renewable Energy Laboratory (NREL, USA) designed the airfoils S8xx for different blade sizes.

Other Universities like Illinois (USA) and Delft (Netherlands) was also contributed to such type of airfoils design.

Du structural reasons, blade root should be of great width in order to increment inertia moment and resist better the moments. By other side, tip blade must have small width because it moves to high velocity. For that, a typical blade airfoil family goes from great width at the root to small width at the tip. Table 3 is indicative about that.

Airfoil	r/R	t/c (%)
tip region	0.95	16
primary outboard	0.75	21
roof region	0.4	24

Table 3. Airfoils with different width along blade wingspan

3.3 Selection criteria of airfoils and rotor blade design

To airfoil election to be used on a rotor blade, we should take in account:

1. From moderate to high relative width (t/c), for a rigid rotor, such relation could be between 16% and 26%. If the rotor were flexible, the relative width should lie between 11% and 16%.
2. 2D Reynolds number of the airfoils should be the appropriate to the design considerations.

Fig. 9. Wind turbines (WT) of different powers, exhibiting aerodynamic different solutions; (top at the left side) WT of 600kW in the wind farm Bajo Hondo in Punta Alta, Argentina, with brake in tip of blade; (top at the right) small WT of 1kW for autonomous supply of relay station of micro waves in surroundings of Puerto Madryn, Argentina; (below) blade of WT of 2 MW in Magdalena's wind farm (Uruguay).

3. Less roughness sensibility is of primary importance due wind turbine stall regulation.
4. Lift budget for minimum drag and C_{lmax}: should be designed for a determinate lift range for minimum drag, despite that for small wind turbines drag acts as passive flow control and have a relative influence upon their function.
5. High lift root airfoils, in order to reduce as much as possible solidity and to increment the starting torque.

For example, Figure 9 show us three wind turbines under operation in different places, for small, medium and high power and, consequently, with different aerodynamic solutions. The first two are located in Argentina and the last one in Uruguay.

Also should be taken in account:

1. Low lift values implies greater solidity and aerodynamic loads.
2. Extreme aerodynamic loads are particularly important for great wind turbines.
3. Low lift airfoils have a smooth stall, which is dynamically positive and the power peaks are reduced.
4. High lift implies small chord values and low operational Reynolds numbers, with subsequent building difficulties.
5. Reynolds number effects are particularly important for small wind turbines.
6. Rotor solidity should be optimal, remembering that´s defined by the relation between blade area and area covered in one complete rotation:

$$\sigma = \frac{Blade\,Area}{\pi \cdot R^2} \tag{34}$$

Low solidity values are related, commonly, with low weight blades and low overall cost

For a given maximum power, optimal solidity depends upon: rotor diameter (big diameter imply low solidity); aerodynamic airfoil (for example, high C_{lmax} imply low solidity); rpm of the rotor (for example, high rpm imply low solidity); blade material (for example, carbon fiber imply low solidity).

4. References

Banks,W.H.H. and Gadd, G.E.; "Delaying effects of rotation on laminar separation". AIAA Journal Vol.1, No.4, Technical note, pp.941-942, 1963.

Carroll C.A., Broeren, A.P., Giguere,P , Gopalarathnam,A , Lyon C.A, and. Selig; M.S; "Low Reynolds Number Airfoil Design and Wind Tunnel Testing"; UIUC, 1990 - 2000

Chandrsuda C. and Bradshaw P.; "Turbulence structure of a reattaching mixing layer". Journal of Fluid Mechanics Vol. 110, pp. 171-194; 1981.

Corrigan, J.J. and Schillings, J.J.; "Empirical Model for Stall Delay due to Rotation". American Helicopter Society Aeromechanics Specialists conf, San Francisco CA, Jan. 1994.

Delnero, Sebastián; "Comportamiento Aerodinámico de Perfiles de Bajo Reynolds, Inmersos en Flujo Turbulento"; Tesis Doctoral, Universidad Nacional La plata; 2007.

Dwyer, H.A. (univ of Calif., Davis) and Mc Crosky,W.J.; "Crossflow and Unsteady Boundary-layer Effects on Rotating Blades". AIAA paper no.70-50, pp.1-15, January 1970.

Eggers, A.J. and Digumarthi, R.; "Approximate Scaling of Rotational Effects of Mean Aerodynamic Moments and Power Generated by the Combined Experiment Rotor Blades Operating in Deep-Stalled Flow". 11-th ASME Wind Energy Symposium, Jan. 1992, pp.33-43.

Eppler, R., and D.M. Somers; "A Computer Program for the Design and Analysis of Low Speed Airfoils"; Technical Memorandum 80210, NASA, Aug. 1980.

Eppler, Richard; "Airfoil Design and Data" . Springer - Verlag , Berlin, 1990.

Gad-el-Hak, M. (2000). "Flow control: Passive, Active and Reactive Flow Management". Cambridge Univ. Press. ISBN 0 521 77006 8.

Glauert, H. (1926). "The Analysis of Experimental Results in the Windmill Brake and Vortex Ring States of an Airscrew", Rept. 1026. Aeronautical Research Committee Reports and Memoranda, London: Her Majesty's Stationery Office.

Glauert, H. 1926. "A General Theory of the Autogyro." ARCR R&M No. 1111.

Glauert, H. 1935. "Airplane Propellers." Aerodynamic Theory (W. F. Durand, ed.), Div. L, Chapter XI. Berlin:Springer Verlag.

Harris, F.D. (Vertol Division, The Boeing Company, Morton, Pennsylvania); "Preliminary study of radial flow effects on rotor blades". Journal of the American Helicopter Society, Vol.11, No.3, pp.1-21, 1966.

Himmelskamp, H. (PhD dissertation, G¨ottingen, 1945); "Profile investigations on a rotating airscrew". MAP Volkenrode, Reports and Translation No.832, Sept. 1947.

Ho C.M. & Huerre P.; "Perturbed free shear layers". Annual Review of Fluid Mechanics 16, pp. 365-424; 1984.

Klimas, Paul C.; "Three-dimensional stall effects". 1-st IEA Symposium on the Aerodynamics of Wind Turbines, London, 1986, pp.80-101.

McGhee, R. J., and W. D. Beasley; "Low - Speed Aerodynamics Characteristics of a 17 - Percent - Thick Section Designed for General Aviation Applications", TN D-7428 NASA, Dec. 1973.

Moriarty P. J., Hansen C. A.,"AeroDyn Theory Manual", NREL EL/-500-36881, December 2005.

Spera, D.A. (Ed.): "Wind Turbine Technology",ASME Press, New York, 1994.

Sørensen, Jens Nørkaer; "A new computational model for predicting 3-D stall on a HAT". 1-st IEA Symposium on the Aerodynamics of Wind Turbines, London 1986.

Sørensen, J.N., Michelsen, J.A., and Schreck, S.; "Navier-Stokes predictions of the NREL Phase-IV rotor in the NASA Ames 80-by-120 wind tunnel". AIAA-2002-0031, ASME Wind Energy Symposium, 2002, pp.94-105.

Tangler, J.L. and Selig, Michael S.; "An evaluation of an empirical model for stall delay due to rotation for HAWTs". In ProceedingsWindpower '97, Austin TX, pp.87-96.

Xu, Guanpeng and Sankar, Lakshmi N.; "Application of a viscous flow methodology to the NREL phase-IV rotor". ASME Wind Energy Symposium, Reno 2002, AIAA-2002-0030, pp.83-93.

Zaman K.B. & Hussain A.K.; "Turbulence suppression in free shear flows by controlled excitation". Journal of Fluid Mechanics 103, pp. 133-159; 1981.

Section 4

Nonstationary Aerodynamics and CFD in Compressible Flow

8

An Approximate Riemann Solver
for Euler Equations

Oscar Falcinelli[1], Sergio Elaskar[1,2],
José Tamagno[1] and Jorge Colman Lerner[3]
[1]Aeronautical Department, FCEFyN,
Nacional University of Cordoba
[2]CONICET
[3]Aeronautical Department, Engineering Faculty,
Nacional University of La Plata
Argentina

1. Introduction

A fundamental subject leading to numerical simulations of Euler equations by the Finite Volume (FV) method, is the calculation of numerical fluxes at cells interfaces. The degree of accuracy of the FV numerical scheme, its ability to capture discontinuities and the correct prediction of the velocity of propagating waves, are all flow properties strongly dependent on the evaluation of numerical fluxes.

In many numerical schemes, the fluxes between cells are computed using truncated series expansions which based only on numerical considerations. These considerations to be strictly valid, must account for some degree of continuity in the functions and in their derivatives, but clearly, these continuity conditions are not satisfied when discontinuous solutions as shock waves or contact surfaces are present in the flow. This type of flow problems nevertheless, were solved with relative success until 1959. In that year Godunov published his work "A finite difference method for the computation of discontinuous solutions of the equations of Fluid Dynamics" (Godunov, 1959) in which an alternative approach for solving the system of Euler equations is presented. This new approach, in striking difference to previous ones, is basically supported by physical considerations and the essential part of it is the so called Riemann solver.

The excellent results obtained with the Godunov technique, prompted several researches to develop new FV numerical schemes for two and three dimensional applications, achieving second order accuracy and total variation diminishing (TVD) properties (Toro, 2009; LeVeque, 2004; Yee, 1989). These new schemes were built around the use of Riemann solvers, making them generally very accurate but computationally expensive. Such high computational cost is attributed to the iterative technique required to solve the system of five nonlinear algebraic equations needed to find, in all cells, an exact solution of the Riemann problem. Alternative schemes computationally less demanding, although less accurate and less robust, were then built based on approximate solutions of the Riemann problem (Toro, 2009).

In this article, a new non-iterative way of solving the full Riemann problem, applicable to three-dimensional and time dependent Euler equations, is presented. This non-iterative solution requires at the beginning of the computation and only once, the generation in situ of tabulated exact Riemann solutions from which the information needed can be retrieved using multiple-linear interpolation. To reduce the time to access during successive calculations, the original five independent variables of the exact Riemann problem are, by way of dimensional analysis, reduced to only three thus allowing the build-up of an easiest to handle data-based matrix with three degrees of freedom.

2. Description of the proposed Riemann Solver

Compressible and inviscid flow problems are governed by the Euler equations. In one dimension (1D), these equations can be written as:

$$U_t + F(U)_x = 0 \tag{1}$$

$$U = \begin{bmatrix} \rho \\ \rho u \\ E \end{bmatrix}; \quad F(U) = \begin{bmatrix} \rho u \\ \rho u^2 + p \\ (E+p)u \end{bmatrix} \tag{2}$$

U is the vector of conservative variables, F the vector of convective flows, t and x are the temporal and spatial coordinates respectively, ρ is the density, u the velocity in the x direction, p the pressure and E the total energy per unit volume. The subscript means differentiation with respect to time and space.

The Riemann problem solves de Euler equations in a 1D domain where initial conditions are given by two different constant states separated by a discontinuity. The solution of this 1D Riemann problem depends directly on the ratio x/t, and it will consist of three types of waves: two nonlinear, shock or expansion fan, and one linearly degenerate, the contact discontinuity. These waves are separating four constant states where the conservative vector U acquires from the left to the right the following values, U_L, U_{L^*}, U_{R^*} and U_R. The subscripts "L" and "R" indicate left and right, respectively, and the symbol "*" identify points located in the state between the nonlinear waves (star region).

To obtain the flow variations produced by the waves, the Riemann invariants technique for expansion and contact waves and of the Rankine-Hugoniot relationship for shock waves, must be implemented. In the Riemann problem, this generally leads to an algebraic system of nine equations with nine unknowns. The unknowns are the velocities of the three waves plus six variables necessary to characterize the states U_{L^*}, U_{R^*}. However, an analysis of the eigenstructure of the Euler equations allows to establish that both pressure p^* and particle velocity u^* between the left and right waves are constant, while the density take the two constant values ρ_{*L} and ρ_{*R} (Toro, 2009). Based on these considerations and after some algebraic calculations, the two equations listed below are obtained and used for solving the Riemann problem.

$$f_L(p_*, U_L) + f_R(p_*, U_R) + \Delta u = 0 \tag{3a}$$

$$u^* = \frac{1}{2}(u_L + u_R) + \frac{1}{2}\left[f_R(p_*,U_R) - f_L(p_*,U_L)\right] \tag{3b}$$

where $\Delta u = u_R - u_L$. Once Eq.(3a) is solved for p_* the solution for u_* is obtained from Eq.(3b) and the remaining unknowns are found by means of standard gas dynamic relations. The functions f_L and f_R represent relations across the left non-linear wave and across the right non-linear wave respectively, and are given by

$$f_K(p_*,U_K) = (p_* - p_K) \cdot \sqrt{\frac{A_K}{p_* + B_K}} \qquad if \qquad p_* \geq p_K \tag{4a}$$

$$f_K(p_*,U_K) = \frac{2 \cdot a_K}{\gamma - 1} \cdot \left[\left(\frac{p_*}{p_K}\right)^{\frac{\gamma-1}{2\cdot\gamma}} - 1\right] \qquad if \qquad p_* < p_K \tag{4b}$$

$$A_K = \frac{2}{(\gamma+1)\cdot\rho_K} \qquad B_K = \frac{(\gamma-1)}{(\gamma+1)}\cdot p_K \tag{5}$$

where γ is the ratio of specific heats, K may be L or R depending on if the flow changes are evaluated across the nonlinear left or right waves, and a_K is the speed of sound in the left or right states. The Eq.(3a) is an implicit algebraic nonlinear equation on the only unknown p_*, and it must be solved using an iterative scheme. Once the pressure in the star region has been obtained, it is possible to calculate by means of explicit expressions the velocity u_* and the density at each side of the contact discontinuity.

In Eq.(3a), neither u_L nor u_R are explicitly written, but only its difference. Therefore the pressure in the star zone becomes only function of five variables:

$$p_* = f_1(\Delta u, \rho_L, p_L, \rho_R, p_R) \tag{6}$$

In this article, the dimensional analysis is used to reduce the number of independent variables necessaries for describing the gas-dynamics Riemann problem. From this point of view, it is possible to consider as reference variables the density and pressure from one side of the Riemann problem, for example, the right side.

$$\rho_{ref} = \rho_R \qquad\qquad p_{ref} = p_R \tag{7}$$

The Eq.(6) represents a relationship between pressures, velocities and densities. By means of the dimensional analysis, the densities can be written in non-dimensional form using a reference value such as ρ_R, the pressures using as reference p_R, and the velocities respect of $(\rho_R\, p_R)^{1/2}$. Then, the solution of a particular Riemann problem is determined by the following three parameters:

$$\pi_1 = \frac{\Delta u}{\sqrt{\dfrac{p_R}{\rho_R}}} \qquad\qquad \pi_2 = \frac{p_L}{p_R} \qquad\qquad \pi_3 = \frac{\rho_L}{\rho_R} \tag{8}$$

Different cases of Riemann problems with identical values of π_1, π_2, π_3 must have similar behavior, and the calculated relation p_*/p_R must have the same value for all cases. To ascertain this behavior for the non-dimensional approach proposed, several numerical examples are analyzed. Seven Riemann problems whose left and right conditions are indicated in the following table, and all of them satisfying the parameters π_1 = -0.78262, π_2 = 50, π_3 = 10, were tested:

ρ_L	u_L	p_L	ρ_R	u_R	p_R
1.2250E+00	1.0000E+02	1.0000E+05	1.2250E-01	0.0000E+00	2.0000E+03
4.9071E+01	8.4770E+02	8.7460E+06	4.9071E+00	6.9994E+02	1.7492E+05
6.7304E+00	6.5231E+02	1.0554E+07	6.7304E-01	2.1402E+02	2.1108E+05
4.1503E+00	7.8027E+02	1.1631E+07	4.1503E-01	1.9437E+02	2.3261E+05
9.4504E+00	6.4262E+02	1.5976E+07	9.4504E-01	1.8756E+02	3.1952E+05
3.0289E+01	2.9038E+02	2.9757E+06	3.0289E+00	1.8067E+02	5.9514E+04
3.6284E+01	3.0129E+02	6.5687E+05	3.6284E+00	2.5420E+02	1.3137E+04

Table 1. Test cases.

The relation p_*/p_R is obtained solving Eq.(3), and for the seven test cases considered it is found p_*/p_R = 13.312, which proves that the proposed non-dimensional analysis works properly. Then, it is possible to write the Eq.(3) only as function of π_1, π_2, π_3 and p_*/p_R.

Since in solving the Eq.(3) there are involved only three independent variables, a data-base matrix with three degrees of freedom containing the values of p_*/p_R for N values of π_1, M of π_2, and Q of π_3 is, in situ generated. Then, to find the solution of a particular Riemann problem, it is only necessary to calculate the corresponding values of π_1, π_2, π_3 and to interpolate for p_*/p_R in the NxMxQ matrix (from now, simply called A-matrix). Finally the pressure in the star zone can be calculated as:

$$p_* = \left(\overline{\frac{p_*}{p_R}} \right) \cdot p_R \tag{9}$$

where $\left(\overline{\dfrac{p_*}{p_R}} \right)$ is the interpolated value from the A-matrix. After calculating the pressure in

the star region, the rest of the variables can be explicitly calculated using the same equations of the exact solver (Toro, 2009).

The previously described procedure involves the use of an interpolated value of pressure to calculate the density and velocity changes across each wave; however it is not the only possible procedure. Others alternatives are for instance, to develop arrangements with dimensionless density or dimensionless velocity variations across each wave.

To increase the approximate solution accuracy, it is desirable that the variation range of the parameters π_1, π_2, π_3 be as small as possible. One way is to reduce the range of π_2 or π_3 avoiding unneeded storage of data in symmetrical Riemann problems.

Two Riemann problems will be symmetric if the following conditions are satisfied:

$$(p_R)_A = (p_L)_B \qquad (\rho_R)_A = (\rho_L)_B \qquad \Delta u_A = \Delta u_B \qquad (10)$$

Unnecessary storage in symmetrical cases can be avoided if it is adopted as the selection criterion of the reference variables (p_{ref}, ρ_{ref}) in Eq.(7), not the left or right pressure and density, but those of the higher pressure side. Thus, the reference state is the higher pressure initial state and the π_2 value will always be less than or equal to one.

To give a physical sense to the non-dimensional π_1 variable it is convenient to re-define it as:

$$\pi_1 = \frac{\Delta u}{\sqrt{\gamma \cdot \dfrac{p_{ref}}{\rho_{ref}}}} = \frac{\Delta u}{a_{ref}} \qquad (11)$$

where a_{ref} is the sound velocity in the reference state and π_1 would be like a Mach number change between the left and right states in the Riemann problem. However π_1 is not strictly a change of the Mach number because a_{ref} is not the sound velocity neither that of the left state or that of the right.

Finally, it is clear that no interpolation is necessary when the solution of the Riemann problem possesses left and right expansion waves, because in this case the Eq.(3) can be solved analytically.

3. Comparison with other Riemann Solvers

To analyze the accuracy and computational efficiency of the proposed Riemann solver, comparisons with others solvers available in the literature (Toro, 2009), are made. Three were selected: one that, iteratively searches for the exact solution of Eq.(3), and others two, that they try to solve the Riemann problem with approximate schemes. To build-up the comparison, Riemann problems were generated randomly with values of the parameters π_1, π_2 and π_3 ranging between the limits set below:

$$-10.05 \leq \pi_1 \leq 4.95 \quad 0.05 \leq \pi_2 \leq 1 \quad 0.05 \leq \pi_3 \leq 5.05 \qquad (12)$$

The following sub-sections explain how each one of the solvers selected for comparison works.

3.1 Iterative Riemann Solver

This solver searches by means of an iterative process the solution of Eq.(3a). This equation for any value of p shall be written as:

$$f_L(p,U_L) + f_R(p,U_R) + \Delta u = R(p) \qquad (13)$$

where $R(p)$ is the residual to cancel; f_L and f_R, are calculated according to Eqs.(4 and 5).

Usually to iteratively solve Eq.(13), the Newton-Raphson method is implemented. This method requires the calculations of the function as well as of its derivative, which should increase the computational cost of the method, but this increase is not significant because the evaluation of the derivative demands simple computations once the function has been evaluated. Since the derivative of the residual function, Eq.(13), is calculated by means of the functions f_L and f_R without considering U_R and U_L that are constant, it can then be written as:

$$\frac{d\left[f_L\left(p,U_L\right)\right]}{dp} + \frac{d\left[f_R\left(p,U_R\right)\right]}{dp} = \frac{d\left[R\left(p\right)\right]}{dp} \qquad (14)$$

The derivatives $\dfrac{d\left[f_L\left(p,U_L\right)\right]}{dp} = f_L'$ y $\dfrac{d\left[f_R\left(p,U_R\right)\right]}{dp} = f_R'$ are calculated as:

$$f_K' = \begin{cases} \sqrt{\dfrac{A_K}{B_K + p}} \cdot \left(1 - \dfrac{p - p_K}{2 \cdot \left(B_K + p\right)}\right) & \text{if} \quad p \geq p_K \\[4ex] \dfrac{1}{\rho_K \cdot a_K} \cdot \left(\dfrac{p}{p_K}\right)^{\frac{-(\gamma+1)}{2 \cdot \gamma}} & \text{if} \quad p < p_K \end{cases} \qquad (15)$$

From the computational point of view it is noted that the most expensive steps involved in the numerical evaluation of Eq.(15) are the powers with fractional exponents. However, Eqs.(4a and 4b) and Eq.(15) show that these computational steps have already been made when the residual $R(p)$ is computed, and as a result of this its derivative calculation is relatively fast.

The Newton-Raphson algorithm applied to Eq.(13) can be written as:

$$p_{i+1} = p_i - \frac{R_{(p_i)}}{R'_{(p_i)}} \qquad (16)$$

where p_i and p_{i+1} are the pressure for the iteration i and $i+1$ respectively, R and R' are the residual function and its derivative.

In solving iteratively Eq.(13) difficulties may appears because, as Eq.(16) suggests, the pressure can becomes negative. To avoid this problem, the residual function is evaluated (considering $p = p_{min}$ and $p = p_{max}$, where $p_{min} = min[p_L, p_R]$ and $p_{max} = max[p_L, p_R]$. If both residuals are positive, the pressure which cancels the residual is less than p_L and p_R, and the Riemann problem has an explicit solution consisting of two rarefaction waves. If the residual corresponding to the maximum pressure is greater than zero and the corresponding to minimum pressure is less than zero, the Riemann problem have as solutions a shock wave and a rarefaction fan, and it is adopted as a first iteration the value of the minimum pressure (which will undoubtedly be lower than the pressure to cancel the residual). When both residuals are negative, the Riemann problem has two shock waves and the sought pressure will be greater than both, so as first iteration the maximum pressure value is adopted.

Using the above described procedure, it is possible to know beforehand the kind of solution expected for the Riemann problem at each iterations, which allows an effective selection of the function given by Eq.(4), and thus slightly shortening the time needed to reach the correct solution.

3.2 Two-Rarefaction Riemann Solver (TRRS)

A particular solution of the gas-dynamics Riemann problem is given when both non-linear waves are rarefaction waves. In this case the pressure in the star region can be obtained analytically.

$$\frac{2 \cdot a_L}{\gamma-1} \cdot \left[\left(\frac{p_*}{p_L} \right)^{\frac{\gamma-1}{2\cdot\gamma}} - 1 \right] + \frac{2 \cdot a_R}{\gamma-1} \cdot \left[\left(\frac{p_*}{p_R} \right)^{\frac{\gamma-1}{2\cdot\gamma}} - 1 \right] + \Delta u = 0$$

$$\Downarrow$$

(17)

$$p_* = \left[\frac{a_L + a_R - \Delta u \cdot \frac{(\gamma-1)}{2}}{\dfrac{a_L}{p_L^{\left(\frac{\gamma-1}{2\cdot\gamma}\right)}} + \dfrac{a_R}{p_R^{\left(\frac{\gamma-1}{2\cdot\gamma}\right)}}} \right]^{\left(\frac{2\cdot\gamma}{\gamma-1}\right)}$$

When the pressure exceeds the value of p_{min}, and the Eq.(4a) instead of Eq.(4b) to calculate the f_K functions is used, there are no discontinuities in the residual function and in its derivatives. This particular behavior of the Eq.(4a) and of its first derivative implies that the error incurred is at most of second order if instead of Eq.(4a), the Eq.(4b) is utilized.

However, if the Eq.(4b) is used to calculate the function f_K the ability to obtain analytically the value p_*, is lost. The TRRS method, suggests to calculate the pressure in the star region always using the Eq.(17), no matter what kind of Riemann problem is studied. The error of this method will be null for the Riemann problems with two rarefaction waves ($p_* \le p_{min}$) and will increase as p_* becoming higher and moves away from p_{min}.

3.3 Two-Shock Riemann solver (TSRS)

The TSRS is the opposite case of the TRRS. In the TSRS the solution of the Riemann problem is obtained considering that both non-linear waves are shocks. Then the functions f_K are given by the Eq.(4b) and the Eq.(3a) can be written as shown bellow:

$$(p_* - p_L) \cdot \sqrt{\frac{\dfrac{2}{(\gamma+1)\cdot\rho_L}}{p_* + \dfrac{(\gamma-1)}{(\gamma+1)} \cdot p_L}} + (p_* - p_R) \cdot \sqrt{\frac{\dfrac{2}{(\gamma+1)\cdot\rho_R}}{p_* + \dfrac{(\gamma-1)}{(\gamma+1)} \cdot p_R}} + \Delta u = 0$$

$$\Downarrow$$

(18)

$$(p_* - p_L) \cdot g(p_*, U_L) + (p_* - p_R) \cdot g(p_*, U_R) + \Delta u = 0$$

where

$$g(p,U_K) = \sqrt{\frac{\dfrac{2}{(\gamma+1)\cdot\rho_K}}{p + \dfrac{(\gamma-1)}{(\gamma+1)}\cdot p_K}} \tag{19}$$

The TSRS does not provide an analytical solution for p_*, and it is necessary an iterative process beginning with an approximation for p_* called p_0. p_0 is used to calculate the value of $g(p_0,U_L)$ and $g(p_0,U_R)$. Then, assuming $g(p_0,U_L)$ and $g(p_0,U_R)$ constants the Eq.(18) is lineal and it is possible to obtain p_* as:

$$p_* = \frac{p_L\cdot g(p_0,U_L) + p_R\cdot g(p_0,U_R) - \Delta u}{g(p_0,U_L) + g(p_0,U_R)} \tag{20}$$

Following the book (Toro, 2009) the approximation of p_* can be expressed as:

$$p_0 = \frac{p_L + p_R}{2} + \frac{u_L + u_R}{2}\cdot\frac{\rho_L + \rho_R}{2}\cdot\frac{a_L + a_R}{2} \tag{21}$$

Although the TSRS scheme, even when applied to cases with two shock waves does not provide an exact solution, it is very robust and has become one of the more implemented approximated schemes to solve the Riemann problem.

3.4 Approximate and Adaptive Riemann solver using the TSRS and TRRS

The adaptive solver developed in this work is compared with an approximate Riemann solver that, as shown by Toro (2009), bind together the advantages of the TRRS and TSRS schemes. This Adaptive Riemann solver obtains the approximate p_0 as given by Eq.(21), then compare this pressure with p_{min} and p_{max} , and the pressure in the star region will be:

$$p_* = \begin{cases} p_{*TRRS} & \text{if } p_0 \leq p_{min} \\[2mm] p_0 & \text{if } p_{min} < p_0 < p_{máx} \\[2mm] p_{*TSRS} & \text{if } p_0 \geq p_{máx} \end{cases} \tag{22}$$

being p_{*TRRS} and p_{*TSRS} the pressures obtained using the TRRS and TSRS solvers respectively.

The comparatives results are presented in Section 5 of this chapter.

4. Description of the implemented schemes

In order to test the usefulness of the proposed Riemann solver, four computer codes were developed based on broadly well-known formulations. Three of them solve one dimensional problem applying second order accurates TVD schemes, and the other one applies a first order method to solve two dimensional problems. These formulations were selected because they require that a Riemann solver be implemented.

To compare the impact of the proposed Riemann solver in terms of accuracy as well as computational efficiency, all the numerical codes built have two versions, one of them works with the exact iterative solver and the other with the one here proposed.

The numerical schemes that the developed codes use are: the two-dimensional Godunov approach, the one-dimensional versions of the HLLC - Harten, Lax, van Leer Contact - (Toro, 2009), MUSCL - Monotonic Upstream-Centered Scheme for Conservation Laws - (van Leer, 1985 and Toro, 1994) and finally the RCM - Random Choice Method - method (Chorin, 1977).

The following sub-sections explain each one of them.

4.1 One-dimensional HLLC TVD method

Consider Figure 1, were the complete structure of the solution of the Riemann problem in terms of the slowest S_L and fastest S_R waves, and a middle wave of speed S_* is contained. Note that the HLLC Riemann solver does not compute the speed of the waves, but in order to determine completely the numerical fluxes an algorithm for computing the wave speeds has to be provided.

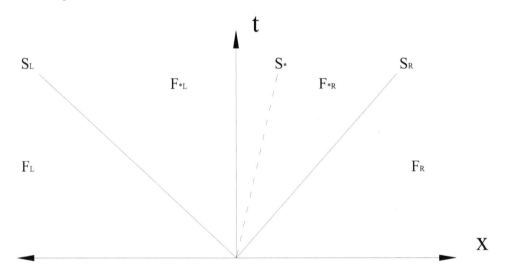

Fig. 1. Riemann problem for the HLLC method.

By applying Rankine-Hugoniot conditions across each of the waves S_L , S_* , S_R it can be obtained

$$R.H. \Rightarrow F_{*_L} - F_L = S_L \cdot (U_{*_L} - U_L) \tag{23a}$$

$$R.H. \Rightarrow F_{*_R} - F_{*_L} = S_* \cdot (U_{*_R} - U_{*_L}) \tag{23b}$$

$$R.H. \Rightarrow F_R - F_{*_R} = S_R \cdot (U_R - U_{*_R}) \tag{23c}$$

These are three equations for the four unknowns vectors U_{*L}, U_{*R}, F_{*L} and F_{*R}. The aim is to find the vectors U_{*L} and U_{*R}, so that the fluxes F_{*L} and F_{*R} can be determined from Eq. (23).

From Eq.(23a) and Eq.(23c), F_L and F_R can be written as:

$$F_L = F_L + S_L \cdot (U_L - U_L) \tag{24}$$

$$F_R = F_R - S_R \cdot (U_R - U_R) \tag{25}$$

Introducing these last two expressions in Eq.(23b), it can be re-arranged as:

$$U_{*R} \cdot (S_R - S_*) + U_{*L} \cdot (S_* - S_L) = F_L - F_R + S_R \cdot U_R - S_L \cdot U_L \tag{26}$$

The Eq.(26) has three scalar equations and six unknowns, the components of U_{*L} and U_{*R}. The following new conditions are now imposed on the approximate Riemann solver

$$u_{*R} = u_{*L} = u_* \quad p_{*R} = p_{*L} = p_* \tag{27a}$$

which are satisfied by the exact solution. In addition, it is justified and convenient, to set

$$S_* = u_* \tag{27b}$$

that is, the star zone velocity must be equal to the contact discontinuity velocity.

Then, only one closure conditions remains to be set. In (Toro, 2009), the following is proposed:

$$\rho_{*K} = \rho_K \cdot \left(\frac{S_K - u_K}{S_K - S_*} \right) \tag{28}$$

Equation (27a) sets that the star zone velocity must be equal to the contact discontinuity velocity. To avoid confusion are called RHS1, RHS2 and RHS3 the first, second and third scalar components of the RHS vector of Eq.(26). It possible to show that to satisfy simultaneously the Eqs.(26 and 27):

$$S_* = \frac{RHS2}{RHS1} \tag{29}$$

$$p_* = \frac{RHS3 - \dfrac{S_*}{2} \cdot RHS2}{(S_R - S_L)}(\gamma - 1) \tag{30}$$

To obtain Eq.(30) was used the relation $E = \dfrac{\rho}{2}u^2 + \dfrac{p}{\gamma - 1}$.

The complete scheme shall consist on obtaining an estimation of S_L and S_R, on calculating S_* and p_* using Eqs.(29 and 30) respectively, and then determining ρ_{*R} and ρ_{*L} through Eq. (27). Finally the flow vectors F_{*L} and F_{*R} by means of the Eq. (24 and 25) are evaluated.

To estimate, either S_L or S_R when there is a rarefaction fan, it has been proposed to use the wave velocity in contact with the undisturbed state, and when there is a shock wave directly to use the shock velocity (Toro, 2009).

The difference between a first order, second order and TVD schemes are inherent to the structure of the numerical fluxes at cells interfaces. Calling wave-1, wave-2 and wave-3 those that separate the L and $*L$, $*L$ and $*R$, $*R$ and R states respectively; the algorithm in this paper implemented is

$$F_{i+1/2} = \frac{1}{2}\left(F_i + F_{i+1}\right) - \frac{1}{2}\sum_{j=1}^{3} sign\left(S_j\right)\phi_{i+1/2}^j \, \Delta F_{i+1/2}^j \tag{31}$$

being i and $i+1$ the left and right cells and $\phi_{i+1/2}^j$ is the limiter function for the wave-j. In this work are shown only the results obtained using a Van Leer limiter function; however the scheme works efficiently with other limiters.

Finally, the flow state vector is actualized at each time step by means of the explicit scheme:

$$U_i^{n+1} = U_i^{n+1} + \frac{\Delta t}{\Delta x}\left(F_{i-1/2} - F_{i+1/2}\right) \tag{32}$$

4.2 MUSCL TVD one-dimensional method

To construct discrete second-order accurate schemes the MUSCL method proposed by Hancock (van Leer, 1985) carries out the following steps:

Step 1. Choice of a suitably slope vector Δ_i and data reconstruction with boundary interpolated values.

Step 2. For each cell the boundary extrapolated values are evolved by a half time interval

Step 3. Solve the Riemann problem with data provided after Step 2.

Step 4. Compute new inter-cell fluxes and state vectors to complete one time interval

In the first step, using the flow solution from the previous time and applying some particular criterion, determine the slope on each cell, as shown in Figure 2.

Next, considering the slope in each cell, the state vector or independent state variables are extrapolated from the cell center to the cell boundary, namely

$$U_i^L = U_i - \frac{\Delta_i}{2} \quad U_i^R = U_i + \frac{\Delta_i}{2} \tag{33}$$

In the second step, it is calculated the evolution of states variables by a time ½ Δt (Fig. 3) according to:

$$\Delta_i^{\frac{\Delta t}{2}} = \frac{\Delta t}{2 \cdot \Delta x} \cdot \left[F\left(U_i^L\right) - F\left(U_i^R\right)\right] \tag{34}$$

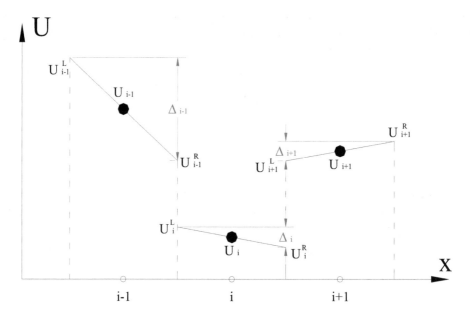

Fig. 2. MUSCL method data reconstruction.

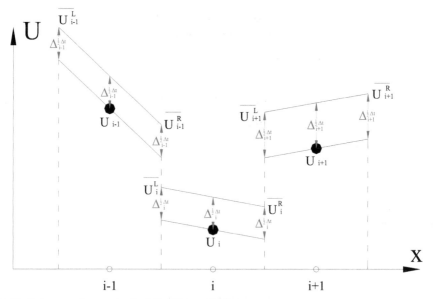

Fig. 3. Half time evolution in the MUSCL method.

Notice, that after using Eq.(34) the discontinuities between cells are actualized, and the Riemann problems are calculated using the half time evolved state vectors. Thus, at the $i+\frac{1}{2}$ interface the right adjacent initial state for the Riemann problem is given by

$$\overline{U_i^R} = U_i^R + \Delta_i^{\Delta t/2} = U_i + \frac{\Delta_i}{2} + \frac{\Delta t}{2\Delta x} \cdot \left[F\left(U_i^L\right) - F\left(U_i^R\right)\right] \tag{35}$$

and the left state by

$$\overline{U_{i+1}^L} = U_{i+1}^L + \Delta_{i+1}^{\Delta t/2} = U_{i+1} - \frac{\Delta_{i+1}}{2} + \frac{\Delta t}{2\Delta x} \cdot \left[F\left(U_{i+1}^L\right) - F\left(U_{i+1}^R\right)\right] \tag{36}$$

Once solved the Riemann problems the new inter-cell fluxes and state vectors are calculated for the complete time step (step four).

In this chapter, the following criterion to evaluate the slopes at the cells is used (Toro, 2009).

$$\Delta_i = \begin{cases} \max\left[0, \min\left(\beta\Delta_{i-1/2}, \Delta_{i+1/2}\right), \min\left(\Delta_{i-1/2}, \beta\Delta_{i+1/2}\right)\right] & \text{si } \Delta_{i+1/2} > 0 \\[2ex] \min\left[0, \max\left(\beta\Delta_{i-1/2}, \Delta_{i+1/2}\right), \max\left(\Delta_{i-1/2}, \beta\Delta_{i+1/2}\right)\right] & \text{si } \Delta_{i+1/2} \leq 0 \end{cases} \tag{37}$$

where:

$$\Delta_{i-1/2} = U_i - U_{i-1} \quad \Delta_{i+1/2} = U_{i+1} - U_i \tag{38}$$

and β is a variable specifying the limiter function. Using Eqs.(37 and 38) the MUSCL scheme becomes TVD.

The results presented in Section 5 has been obtained using the SUPERBEE limiter function ($\beta = 2$).

4.3 The TVD Random Choice Method (RCM)

The RCM method can be seen as a modification of the Godunov method (Chorin, 1977). In the Godunov method, at the beginning of each time step, the state vector or independent states variables are considered constant within each cell. This piece-wise constant distribution of data at each time level, define local Riemann problems at the interface between neighboring cells.

To advance to the next time level, the Godunov method utilizes an integral average of local solutions of Riemann problems. Then, new averaged state vectors for all the cells are calculated via integrals

$$U_i^{n+1} = \frac{\int_{-\Delta x/2}^{\Delta x/2} U_{(x)}\,\delta x}{\Delta x} \tag{39}$$

where $U_{(x)}$ is the state vector solution within the i-cell, which has been obtained as solution of two adjacent Riemann problems (right $i+\frac{1}{2}$, and left $i-\frac{1}{2}$).

The RCM and Godunov methods are similar in the sense that both use the exact solution of the Riemann problem. However, the RCM, instead of averaging according to Eq.(39), advances to the next time level assigning to each cell a picked value $U_{(x)}$ of the state vector contained in the local solution. The picked up state depends on a point x randomly chosen within the sampling range $-\Delta x / 2$ and $\Delta x / 2$.

The RCM main advantage stands in its capacity to capture discontinuities separating constant states. It does not introduce artificial dissipation, shock waves and contact discontinuities are solved with infinite definition: the complete jump is produced in only one cell, and this accuracy does not get lost during time evolution.

However, the scheme has some disadvantages:

-   The scheme introduces discontinuities in zones with smooth variations.
-   The discontinuity velocities are random variables, only their average is the correct velocity. Usually the discontinuity places are not determined correctly.
-   The scheme it is not strictly conservative.
-   The randomness is tolerable when solving homogeneous systems, i.e. no source terms.
-   The RCM can not be applied to solve multidimensional non linear problems via splitting techniques.

4.4 Two-dimensional Godunov method

It is well known that the original scheme (Godunov, 1959) is only first order accurate, which make it unsuitable for application to practical problems. Well resolved simulations will require the use of very fine meshes with the associated computing cost. Therefore, the original scheme was modified by incorporating concepts related to integral formulation of the Euler equations. These concepts have allowed substantial increments of computing time intervals.

The theoretical foundations of updated Godunov schemes are based on characteristics of the Riemann problem solution. This solution depends on the ratio x/t, but it does not on time alone or position alone. At any cell interface ($x = 0$, for the local Riemann problem), the state vector remains constant until a coming wave from the neighbor cell reaches the interface. An integral analysis shows that the cell average state vector U can be calculated analytically using the flux vectors at the cell boundaries. However, it is necessary to solve two Riemann problems at the interfaces of the cell in question. The state vector change ΔU_i in the i-cell and during one time interval Δt can be evaluated as:

$$\Delta U_i = \frac{\Delta t}{\Delta x}\left[F\left(U^{i-1/2}_{x/t=0}\right) - F\left(U^{i+1/2}_{x/t=0}\right)\right] \tag{40}$$

were $F(U_{x/t=0})$ are the Godunov inter-cell numerical fluxes; and $U^{i-1/2}_{x/t=0}$ and $U^{i+1/2}_{x/t=0}$ are the similarity solutions evaluated at $x/t=0$ of the Riemann problems at the interface $i-\frac{1}{2}$ e $i+\frac{1}{2}$ respectively.

The fluid dynamics fundamentals of the Gudonov method applicable to two dimensional problems, are similar to those explained above. However to account for the physical two-dimensionality of non Cartesian geometries, some changes must be introduced. These are:

-   The flux balance must be extended to all sides of two-dimensional cells
-   To solve the Riemann problems, the direction of an outward unit vector normal to each side of multilateral shaped cell has to be determined.

Assuming that the local Riemann problems, in a system of coordinates aligned with the normal unit vector to each side of the cell have been solved, and the corresponding numerical fluxes have been obtained, then the alternative expression to Eq.(40) for non Cartesian geometries becomes

$$\Delta U_i = -\frac{\Delta t}{A}\sum_{j=1}^{k}\overline{F}\left(U_{x/t=0}^{i-1/2}\right)\bullet \overline{n}_j \Delta s_j \tag{41}$$

here k is the numbers of cell sides, and $\overline{F}$ the flux vector. $\overline{n}_j$ is the outward unit normal vector , $\bullet$ expresses the interior product and Δs_j is the side length.

5. Results

Results obtained using the approximate TRRS, TSRS and Adaptive RS, are in the next sub-section presented and compared with the new scheme proposed in this article. Comparison with results produced by HLLC, MUSCL and RC methods are presented, in the following after sub-section.

5.1 Results using approximate Riemann solvers

To analyze the behavior of the approximate solvers with randomly chosen values of the parameters π_i within the range given by Eq.(12), 10^6 cases of Riemann problems are studied. Of them, 65% were cases with two shock waves, 6% with one shock and one rarefaction wave and the remaining 29% with two rarefaction waves.

To systematized the analysis, in all Riemann problems the following initial conditions to the right state are established

$$\rho_R = 1\frac{Kg}{m^3} \qquad u_R = 0\frac{m}{s} \qquad p_R = 1\frac{N}{m^2} \tag{42}$$

The initial conditions for the left states are calculated using values of the parameters π_1, π_2 and π_3 picked-out from the intervals defined in Eq.(12) :

$$u_L = -\pi_1 \cdot \sqrt{\gamma \cdot \frac{p_R}{\rho_R}} \qquad p_L = \pi_2 \cdot p_R \qquad \rho_L = \pi_3 \cdot \rho_R \tag{43}$$

To heighten the behavior of the approximate Riemann solvers, in the following table are listed the worst approximate test values for the TRRS, the TSRS, the adaptive RS and the new proposed scheme.

Solver	ρ_L	u_L	p_L	ρ_R	u_R	p_R	Solver prediction	Exact solution
TRRS	4.9733	11.8082	0.0507	1	0	1	998.7362	81.2775
TSRS	4.9182	11.8582	0.0564	1	0	1	31.8961	81.6784
Adaptive	4.9182	11.8582	0.0564	1	0	1	31.8961	81.6784
Proposed	0.0739	11.8752	0.9274	1	0	1	9.5344	9.6541

Table 2. Worst approximated solutions after solving Riemann test cases.

Note that for the three first solvers, the worst results appear when there are relatively strong shock waves. Also, it is noted that the most poorly test case predicted is the same for the TSRS and the Adaptive Riemann solvers. This is so because when the star pressure is highest that both initial pressures, the Adaptive solver uses the same calculation scheme that the TSRS.

The percent error for each tested Riemann solvers is:

TRRS	TSRS	Adaptive	New scheme
1128.80%	60.95%	60.95%	1.24%

Table 3. Percent error for Riemann solvers.

The CPU time used by each approximate solver given as a percentage of the necessary CPU time for the exact Riemann solver is:

TRRS	TSRS	Adaptive	New scheme
38.62%	25.93%	31.60%	34.57%

Table 4. Percent of CPU time.

It is probable that the comparison between Riemann solvers based only on worst test results may not be considered representative and in consequence objectionable. Therefore, another variable based on the average error of all the pressures computed in the star region for the approximate solvers, is introduced. The results (in N/m^2) are shown below:

TRRS	TSRS	Adaptive	New scheme
37.371	7.1504	7.0931	0.0019

Table 5. Average error.

Repeating the previous analysis, but adding the restriction that the exact pressure p^* in each case is bounded to $0.1\ p^L < p^* < 10\ p^L$ and $0.1\ p^R < p^* < 10\ p^R$, and in this manner avoiding the formation of high-intensity shocks and near-vacuum conditions, the worst computed cases for each solver are:

Solver	ρ_L	u_L	p_L	ρ_R	u_R	p_R	Solver prediction	Exact solution
TRRS	2.4112	4.2348	0.9999	1	0	1	12.5554	9.9950
TSRS	0.8759	5.3169	1.0013	1	0	1	6.7847	9.9618
Adaptive	0.8759	5.3169	1.0013	1	0	1	6.7847	9.9618
Proposed	0.0712	11.7759	0.9424	1	0	1	9.2260	9.3234

Table 6. Worst approximated solutions after solving Riemann test cases with restrictions in p^*.

The percent error for each of the tested Riemann solvers using a bounded p^* are:

TRRS	TSRS	Adaptive	New scheme
25.62%	31.89%	31.89%	1.05%

Table 7. Percent errors.

5.2 Results obtained using different numerical schemes

To compare the goodness and shortcoming of the HLLC, MUSCL, RCM and two-dimensional Godunov numerical schemes, an identical test case for all of them is implemented. The two-dimensional software inclusive, it is compared solving the same one dimensional test case.

The selected test case is a 2 meters length shock tube (and for the two-dimensional simulations $0.067m$ in height). The shock tube was selected because it is possible to reach an exact solution, and is a very popular benchmark for compressible computational fluid dynamics.

The shock tube has two sections of equal length separated by a diaphragm (discontinuity on the initial condition), and both sections are filled with air at the same temperature. Initially, the velocities along the tube are null. Inside the right section, the initials pressure and density are $p = 10^5 N / m^2$, $\rho = 1.225 kg / m^3$ respectively, and in the left section are $p = 10^4 N / m^2$, $\rho = 0.1225 kg / m^3$. For these initial conditions, the solution after the diaphragm is broken is known. It is composed by a shock wave traveling to the right at $543.4 m/s$, and one contact surface also moving to the right at $277.6 m/s$. There is also, a rarefaction fan traveling to the left, its wave tail is moving at $338.1 m/s$ (the sound speed of the stagnant gas in the left section) and its front at $4.9 m/s$.

The flow properties at the four states limited by the above described discontinuities and the continuous wave are:

	L	*L	*R	R
ρ	1.225	0.4995	0.2504	0.1225
u	0	277.6	277.6	0
p	100000	28482	28482	10000

Table 8. Shock tube states flow properties.

In all tests the mesh has 200 cells evenly distributed along the tube. In the two-dimensional simulations a structured mesh possessing 2400 triangles is used (see Figure 4)

A

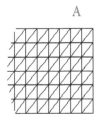

Fig. 4. Mesh for two-dimensional simulations.

The obtained results are presented in Figures 5 to 8. In each figure there are three lines; one represents the exact solution (blue), another the solution using the exact Riemann solver

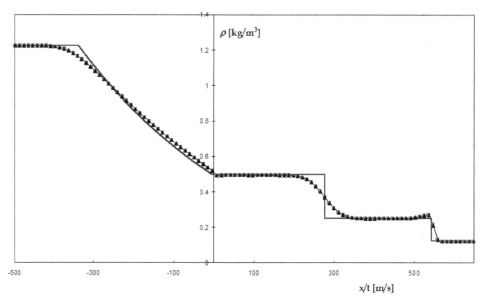

Fig. 5. Two-dimensional Godunov method. Blue line: theoretical solution. Red squares: exact solvers. Black triangles: approximate solver.

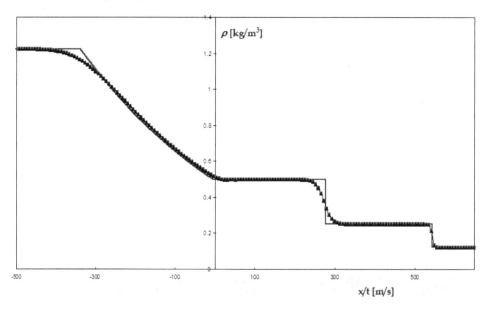

Fig. 6. One-dimensional HLLC method. Blue line: theoretical solution. Red squares: exact solvers. Black triangles: approximate solver

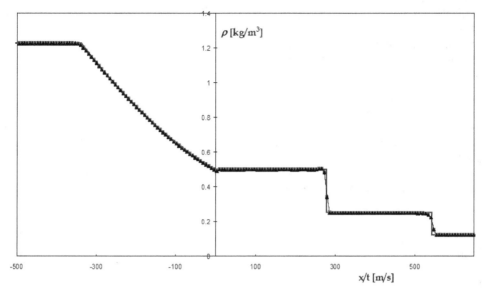

Fig. 7. One-dimensional MUSCL method. Blue line: theoretical solution. Red squares: exact solvers. Black triangles: approximate solver

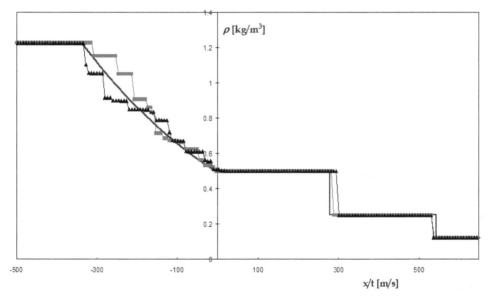

Fig. 8. One-dimensional RCM method. Blue line: theoretical solution. Red squares: exact solvers. Black triangles: approximate solver

(red with squares), and finally the third line shows the results obtained using the approximated Riemann solver proposed in this work (black with triangles). The two dimensional Godunov results shown are computed values along the tube center line.

From Figures 5 to 8 it can be deduced that the results calculated using, either the exact Riemann solver or the new proposed approximation, are practically identical, except for the RCM method. The differences between the numerical results presented in Figures 5 to 7 are less than 1%. For the RCM, the differences are greater; however it is believe that such differences are mainly due to the method randomness, and not to the Riemann solver itself.

In Figure 9 are plotted percentages of the computing time spent for each of the considered numerical schemes to complete a determined percent of the given task, either using the exact Riemann solver or the proposed approximation. Note that in Figure 9, the one hundred percent value has been assigned to all the results that the numerical schemes have produced using the new approximation in solving the Riemann Problem, while the other ones also plotted in percentages, are obtained through the same numerical schemes, but using now the exact Riemann solver.

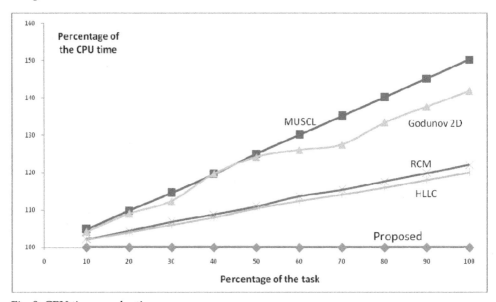

Fig. 9. CPU times evaluation.

It can be seen that using the proposed new approximate Riemann solver, less computing time is needed. In terms of CPU time, the percentages of savings achieved are listed bellow

GODUNOV 2D	HLLC	MUSCL	RCM
41.8%	20.0%	50.2%	22.1%

Table 9. Percent of CPU time.

6. Conclusions

After comparing the accuracy of the new proposed non iterative Riemann solver with the TRRS, the TSRS and the Adaptive RS, it has been found that for all pressure values computed in the star region, the average error of the new solver is notably smaller than the average error of all the others approximated Riemann solvers (Table 5). In terms of the worst pressure values obtained in the star region (Table 2), the percentage error of the new solver is almost 50 time smaller than the best percentage obtained by the other solvers (Table 3). From the point of view of computational cost, it is higher if compared with the Adaptive solver and TSRS by 33.3 % and 9.4% respectively, but lower if the comparison is made with the TRRS. However, the new solver presented has a higher cost-benefit ratio.

It can be argued that the extremely high degree of uncertainty presented by the TRRS was due to the range of sampled Riemann problems (they mostly were problems with two shocks). However, in cases with well defined two rarefaction waves, the TRRS does not offer advantages over an exact Riemann solver because both perform the same operations.

Although the proposed solver has already shown to be efficient, it still requires a sequence of operations that others solvers do not need. For instance, before carrying out the interpolation on a data-base matrix of Riemann solutions, the matrix must be generated (a 100x100x100 matrix like the one used in this paper, requires the exact solution of 106 Riemann problems). However, this matrix is calculated only once at the beginning of the computation and it has only three degrees of freedom, which makes it of easy handling.

The numerical results obtained with HLLC, MUSCL and Godunov schemes, have shown that the new solver is accurate and robust, and no significant differences were found when results are compared with the exact Riemann solver. In addition, it shows appreciable advantages in terms of CPU time.

Modifications on the implementation of the new solver are suggested to benefits from the advantage of using the fully analytical solution in two expansion waves Riemann problems, and only the new approximate solver when there are Riemann problems with shock waves. Furthermore, it shall be desirable to redefine the parameter π_1 avoiding the square root, since it is the more expensive numerical operation.

7. References

Chorin, A. (1977). Random Choise Solutions of Hyperbolic Systems. *Journal of Computational Physics*, Vol. 22, pp. 517-272, ISSN 0021-9991.
Godunov, S. (1959). A Finite Difference Method for the computation of Discontinuous Solutions of the Equations of Fluid Dynamics. *Mat. Sb.*, Vol. 47, pp. 357-393 (In Russian).
Leveque, R. (2004). *Finite Volume Methods for Hyperbolic Problems* (Second Edition), Cambridge University Press, ISBN 0-521-00924-3, Cambridge.
Toro, E., Spruse, M. and Speares, W. (1994). Restoration of the Contact Surface in the HLL-Riemann solver. *Shock Waves*, Vol. 4, pp. 25-34, ISSN 0938-1287.
Toro, E. (2009). *Riemann Solvers and Numerical Methods for Fluid Mechanics. A practical introduction* (Third Edition), Springer-Verlag, ISBN 978-3-540-25202-3, Berlin.
van Leer, B. (1985). On the Relation Between the Upwind-Differencing Schemes of Godunov, Enguist-Osher and Roe. *SIAM Journal of Scientific Computing*, Vol. 5, No. 1, pp. 1-20, ISSN 1064-8275.

Yee, H. (1989) *A Class of High Resolution Explicit and Implicit Shock-Capturing Methods.* NASA Technical memorandum 101088. Ames Research Center, California.

9

Aerodynamic Performance of the Flapping Wing

Abbas Ebrahimi and Karim Mazaheri
Aerospace Systems Center of Excellence,
Sharif University of Technology
Iran

1. Introduction

Flapping is described as simultaneous heave and pitch oscillations of a wing. The dominant way for producing lifting and propulsive force in natural flight is flapping (Pennycuick, 2008; Shyy et al., 1999). Unlike the most common aerial vehicles, in flapping flight, both lift and thrust forces are produced simultaneously by flapping wings. Apparently, flapping is an efficient way for flight in low Reynolds numbers. Currently, the field of low Reynolds number aerodynamics is receiving considerable attention because of a global recent interest in development of Micro Air Vehicles (Shyy et al., 2008).

However, the aerodynamics of flapping flight is not fully understood at present and for the same reason, no design method for flapping wings is readily available (Shyy et al., 1999). Both experimental works (Murphy, 2008; Wilson & Wereley, 2007; Hong & Altman, 2007; Isaac et al., 2006; Muniappan et al., 2005; Ames, 2001) and numerical studies (Mantia & Dabnichki, 2009; Fritz & Long, 2004; Ho, 2003; Jones et al., 2002; Smith et al., 1996; Vest & Katz, 1996) are carried out in this field to explore more about the complex aerodynamic phenomena occurring during the flapping flight. The experimental works include quantitative studies like measuring the forces and moments on a flapping wing in a wind tunnel and qualitative studies such as visualizing the flow in the wakes behind a flapping wing. Although the findings of these studies generate deep insight into the flow structure around flapping wings, this information is generally very qualitative and there is no straight way to use this information for designing flapping wing vehicles.

Ref. (Lasek & Sibliski, 2003) describes the nonlinear modeling of the six degrees of freedom motion of a flapping wing Micro Aerial Vehicle in a turbulent atmosphere. Gebert and Gallmeier (2002) also have investigated the dynamic equations of the motion of flapping wing vehicles. They have found a system of equations, but they have not analyzed these equations to see the effects of different parameters on overall flight characteristics, or on temporal flight dynamics. In (Willis et al., 2006), computational schemes are used to optimize the flight power of a flapping wing. The parameters considered there are: the flapping frequency, the flapping amplitude, and the addition of a mid-wing hinge for articulated flapping flight. The flight power coefficient is minimized for specified flight conditions over a restricted set of flapping parameters.

In Gallivan and DeLaurier (2007), an experimental aerodynamic performance investigation was done on a single set flapping membrane wings with different wings aspect ratio, weight, spar rigidity and batten tailoring. Aditya and Malolan (2007) showed that the Strouhal number affects the propulsive thrust force of a flapping wing. The optimum value is corresponding to a narrow range of Strouhal numbers, which is recommended to be used by wing designers. Lin et al. (2006) also measured the aerodynamic forces for a flexible membrane flapping wing for different flapping wing frequencies, angles of incidence and wind speeds. They observed that the lift force will increase with the increase of the flapping frequency under the corresponding flying speed, while a decrease in the angle of incidence in the same flapping frequency will result in flying speed increase, and a slight decrease in the lift force.

In (Kim et al., 2008), the aerodynamic characteristics due to the effects of the camber and the chordwise wing flexibility were investigated. The experimental results demonstrate that the effect of the camber generated by the macro-fiber composite (MFC) produces sufficient aerodynamic benefit. The chordwise wing flexibility is one of the important parameters affecting to the aerodynamic performance, and can help the wing to stabilize the small leading edge vortex in the unsteady flow condition.

In another experimental investigation, Pfeiffer et al. (2010) proposed the time-efficient integrative simulation framework for the trimmed longitudinal flight of model ornithopter based on the flexible multi-body dynamics and considering fluid-structure interaction. They observed a limit-cycle-oscillation of flight state variables, such as pitch attitude, altitude, flight speed, during the trimmed flight of the model ornithopter. The concept of the "zero moment point" is introduced to explain the physics of ornithopter trimmed longitudinal flight.

In (Mazaheri & Ebrahimi, 2010a), we have experimentally studied the effect of the flexibility on the aerodynamics of flapping wings. The objective of this research is to provide further insight into the flight performance of flapping wing vehicles. The flight envelopes and design curves of flapping wing vehicles are not well developed yet. Our main target here is to formulate a performance analysis procedure for a flapping wing, based on its wing experimental aerodynamic data. A mechanical flapping system capable of producing flapping motion for a special flexible membrane wing is designed and built. The aerodynamic performance of the fabricated wing is experimentally studied through time varying lift and thrust measurements. This experimental work is conducted in a large, low-speed wind tunnel. Results are plotted with respect to the flapping frequency, wind tunnel velocity and angle of incidence. These results are used here to optimize the design of a flapping machine. To do this, we have introduced two flight performance criteria, and have presented a new algorithm to use aerodynamic information to optimize flight conditions according to these criteria. On the basis of this, one finds the optimum value of angle of incidence and flapping frequency based on vehicle weight and speed.

2. Aerodynamic parameters

Here we are going to model the aerodynamic performance of a flapping wing, which may allow us to optimize its design and performance. This may be used at the same time as a design method for design of flapping wing vehicles for different applications or missions.

Let us assume that lift (L) and thrust (T) forces are functions of flapping frequency (f), the wind speed (V), the angle of attack (α), kinematic viscosity of air (v), mean chord (c), and the wing design:

$$L = F(f, V, \alpha, v, c, \text{wing shape}) \qquad (1)$$

$$T = F(f, V, \alpha, v, c, \text{wing shape}) \qquad (2)$$

Note that for our application, viscosity of air is almost constant. The flapping vehicle weight (W) depends on mission, and in cruise flight it is equal to the average lift force ($L = W$). Generally, flapping vehicles experience acceleration or deceleration, but for steady state flight (cruise flight condition), the average force in the flow direction (i.e. thrust minus drag) is zero ($T = 0$), and the wind speed is equal to the cruise velocity ($V = U$). Note that the aerodynamic force in the flow-direction depends on the drag force of the body and the wing itself, and the thrust force generated by the wing. In our analysis, we could not distinguish between the drag and thrust force of the wing, and we could measure or compute the total thrust force, as combination of these two. This is the force that we measure in the wind tunnel. Basically, in flapping vehicles it is not meaningful to distinguish between these two forces. With the above two constraints and using relations (1) and (2), we may find the cruise speed for a given wing design as:

$$\left.\begin{array}{l} L = W = F(f, V, \alpha) \\ T = 0 \ \Rightarrow V = U = F(f, \alpha) \end{array}\right\} \Rightarrow V = U = F(f, \alpha, L = W) \qquad (3)$$

For each flapping frequency and angle of incidence, there is either no or one stable cruise velocity. Similarly, for each cruise speed and flapping frequency, there is at most only one stable angle of incidence.

3. Flapping mechanism and flexible membrane wing

The propulsion system of a flapping wing aircraft consists of different components including a battery, an electric motor, a gearbox, a flapping mechanism, and flexible wings. The flapping wing mechanism is adapted from a Cybird P1 remotely controlled ornithopter (Kim et al., 2003). Figure 1 shows the flapping mechanism and the range of operating angles of the flapping mechanism. The design parameters of flapping mechanism and the analytical results of its kinematics were presented in (Mazaheri & Ebrahimi, 2010b). This simple four-bar crank rocker mechanism transforms the rotational motion of a small electric motor to a harmonic flapping motion. One flapping wing period can be divided into: upstroke and downstroke. The flapping frequency is controlled directly by altering the input voltage. The follower link oscillates between +29° and -21° from the horizontal line. Figure 2 shows analytical results for angular displacement and the corresponding velocity and acceleration of the follower link versus time, for a full period of flapping. During the test, the flapping mechanism is capable of flapping its wing at a maximum frequency of 10 Hz.

Design and development of an efficient wing is the most significant requirement towards making a flapping wing vehicle. Although many researchers have noted that flexible wings play an important role in the aerodynamics of flapping flight, there exist few, although important, studies dealing with flexible rather than rigid wings in flapping flight (Isogai & Harino, 2007; Wu et al., 2008; Singh, 2006; Smith, 1996; Daniel & Combes, 2002).

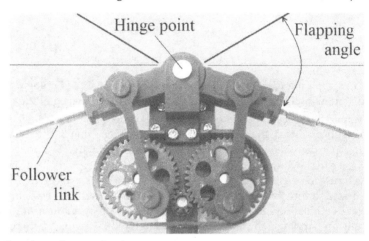

Fig. 1. The flapping wing mechanism

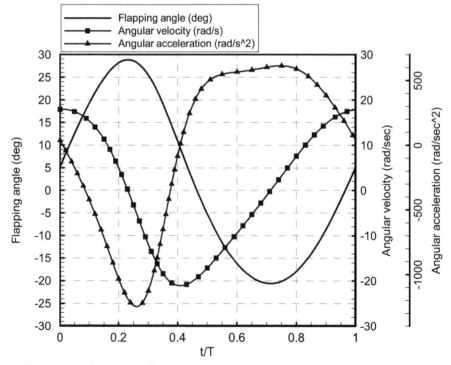

Fig. 2. Flapping angle, angular flapping velocity and angular flapping acceleration for one complete cycle (7 Hz frequency)

It is vital that the aeroelastic similarity parameters be the same for two wings, to have similar aerodynamic performance. The main parameters affecting aeroelastic performance are mass and stiffness distribution.

Figure 3 shows the structure and a planform view of the flexible membrane wing developed here. The wing has a roughly half-elliptical planform with a span of 0.8 m, mean chord of 0.14 m, single wing area of 0.047 m², mass of 14.4 grams and an aspect ratio of 6.8. The main parts of the wing are the spar, the ribs and the skin. The wing's cover is made of nylon sheet and is stiffened by carbon-fiber ribs. As the main spar, one rib makes up the leading edge of the wing, while three other ribs run in the chord direction. The main spar and ribs virtually provide the wing's stiffness. The dynamic bending and twisting of the wing determine the generated lift and thrust. The root of wing's leading edge is rigidly attached to the follower link of the flapping mechanism (see Fig. 1 and 3).

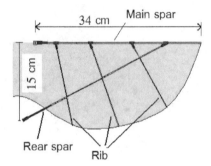

Fig. 3. A planform view of the flexible membrane wing shape schematic

4. Experimental set-up

A test bed, as shown in Fig. 4, is built to investigate the aerodynamic performance of a flapping wing vehicle. To simulate the effects of forward flight, a large closed-loop, low-speed, open test section wind tunnel with a test section of 2.8m x 2.2m and maximum speed of 48 m/sec is used. The velocity nonuniformity in the test section is less than 0.2% and the turbulence intensity in the center of the test section is less than 0.13%. To avoid wall interference effects, we have used an open test section wind tunnel. Figure 5 shows a view of the test bed and the mechanism in the open test section of the wind tunnel.

Unsteady lift and thrust forces are measured using a one-dimensional load cell. The load cell is calibrated statically using known weights and the measured calibration factors are used to convert the gauge voltage signals to forces. For the load cell, the measurement in the range up to 1000 gram shows that the relative error is below 0.5%. Hysteresis, nonrepeatability and nonlinearity of this sensor are below 0.1%. The natural frequency of this sensor (280 Hz) is well above our flapping frequencies. The load cell is installed in the correct direction to measure horizontal and vertical forces respectively. Note that the angle of incidence is measured as the angle between body horizontal axis and the flow direction (Fig. 5) and also, there is 2° positive angle between the flapping axis and the body horizontal axis. The signals received from the strain gauges are amplified electronically using an amplifier before being sampled at 1000 Hz. To measure the wing's power usage, a current measuring unit is designed to use the CSNE151-100 closed loop current sensor. The current sensor is appropriate for variable speed drives and measures AC or DC current with 0.5% accuracy.

All signals are acquired using a data-acquisition board. Then, the data are filtered in an online process using a 3rd order low-pass digital Butterworth filter with a cutoff frequency of about 15 Hz. This cutoff frequency will filter the high-frequency noises generated by motor jitter or vibration of other components. The first natural frequency of the wing is much higher than the flapping frequencies considered here.

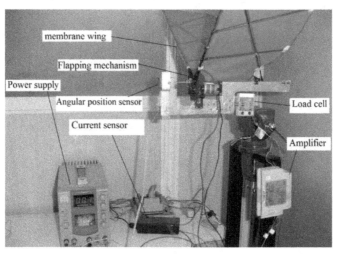

Fig. 4. Experimental set-up

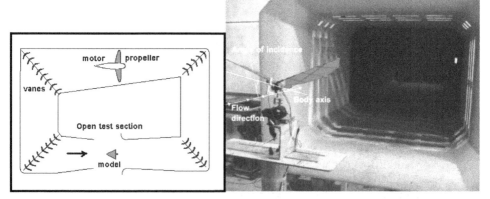

Fig. 5. Flapping wing system in open test section wind tunnel

5. Results and discoussion

5.1 Modeling experimental aerodynamic parameters

Tests were conducted at four different free-stream velocities, between 6 and 12 m/s. For each free-stream velocity, flapping frequency has been varied from zero to 9 Hz. The experiments were repeated for different angles of incidence (zero to 20°). The generated lift, and thrust, and the power consumption are measured in each different case ($T = F(f ; V, \alpha = \alpha_i)$ and $L = F(f ; V, \alpha = \alpha_i)$). Most graphs presented here show the time average of the measured data, based on more than ten cycles.

The average thrust (i.e. thrust minus drag) and lift variation for different free-stream velocities, flapping frequencies and 10° angle of incidence are plotted in Fig. 6. Increase in flapping frequency always result in higher propulsive force. At the same time, increase of wind speed, as expected, results in lower propulsive force (due to increase in drag forces). The general trends of thrust force do not change in different angles of incidence. Also, there is almost no change in lift for low flapping frequencies (up to 5Hz), but it increases for higher flapping frequencies proportional to the flapping frequency. The wind speed also always increases the lift force. The general trends of lift force are similar for another angles of incidence, and that for higher flapping frequencies, the increase rate of lift due to flapping frequency is higher for lower angles of incidence.

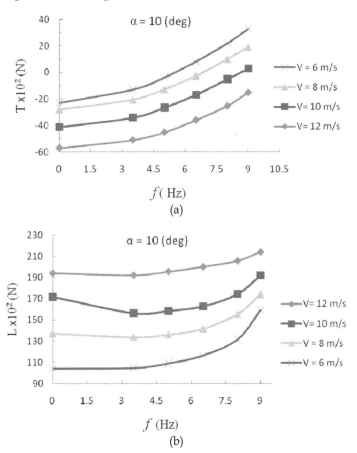

Fig. 6. The average thrust and lift force versus flapping frequencies for different wind speeds and $\alpha = 10°$

5.2 Formulation of flight envelopes

Flight envelopes and design curves are not considered and analyzed by researchers active in this field. Here, we formulate a new scheme to find flight envelopes for a particular flapping wing vehicle. This algorithm is stated in four steps below. The main performance parameter is the possible cruise speed. Since power consumption is basically a function of frequency, for each given weight and payload, one may find the best cruise speed for maximum range, which is corresponding to minimum ratio of power to cruise speed ($\frac{P}{U}$) (i.e., energy usage for unit distance), or for maximum endurance corresponding to minimum power (i.e., minimum flapping frequency).

1. Imposing the no-thrust constraint ($T = 0$) for each angle of incidence, using Fig. 6a, $T = F(f; V, \alpha = \alpha_i)$, one finds possible cruise speed for each flapping frequency, $U = F(f; \alpha = \alpha_i, T = 0)$. Results for cruise speed as a function of the flapping frequency, for various angles of incidence are shown in Fig. 7. Also the power consumption at different flapping frequencies (cruise flight velocities) and angles of incidence (i.e. $P = F(f; \alpha = \alpha_i, T = 0)$ and $P = f(U; \alpha = \alpha_i, T = 0)$) could be derived as shown in Fig. 8. Similarly, one may produce ratio of power to cruise speed criteria, $\frac{P}{U} = F(f; \alpha = \alpha_i, T = 0)$. These results are shown in Fig. 9. We will use this diagram later to find optimum conditions for the best possible range for each given vehicle weight.

2. Use Fig. 6b, $L = F(f; V, \alpha = \alpha_i)$, to find the lift force for cruise. For this cruise flight, the average lift force coefficient as a function of the flapping frequency for various angles of incidence, $L = F(f; \alpha = \alpha_i, T = 0)$, and various flight speeds, $L = F(f; U = U_i, T = 0)$, is shown in a carpet plot in Fig. 10.

3. Now for a given vehicle weight ($L = W_i = 0.15\ kg$), one may use Figs. 7 and 10 to draw cruise speed versus flapping frequency, $U = F(f; L = W_i)$. These results are shown in Figs. 11a and 11b, and values of P, α, and $\frac{P}{U}$ are indicated at each point. Similarly, one may use Figs. 7 and 10 to find angle of incidence versus flapping frequency, $\alpha = F(f; L = W_i)$. Figures 12a and 12b show the results, and the values of P, α, and $\frac{P}{U}$ are indicated at each point.

4. For each point in Figs. 11a and 11b or in Figs. 12a and 12b, one may find values of P, α and $\frac{P}{U}$ and draw two different graphs:

    a. The flight power curve, $P = F(f; L = W_i)$, which helps us to find the maximum endurance conditions for a given vehicle weight.

    b. The flight power to cruise speed curve, $\frac{P}{U} = F(f; L = W_i)$, which helps to find the maximum range conditions for a given vehicle weight.

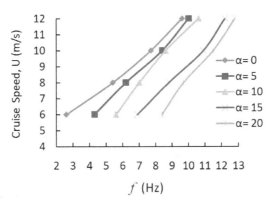

Fig. 7. Cruise speed as a function of the flapping frequency, for various angles of incidence

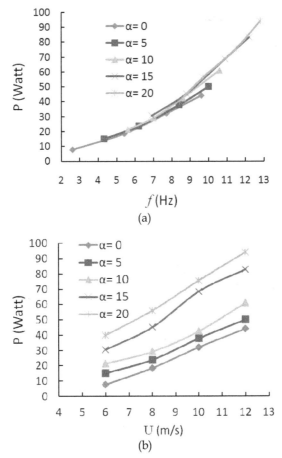

Fig. 8. The input power as a function of the flapping frequency (cruise flight velocities), for various angles of incidence

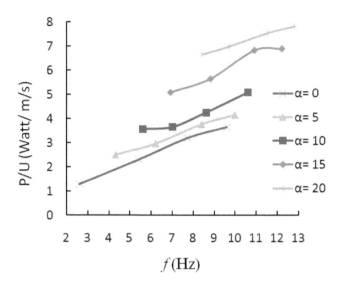

Fig. 9. Energy usage over unit distance as a function of the flapping frequency, for various angles of incidence

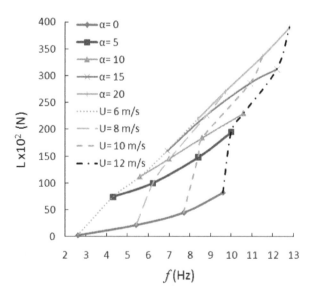

Fig. 10. The average lift as a function of the flapping frequency, for various angles of incidence ($L = F(f\,; \alpha = \alpha_i, T = 0)$), and various cruise speeds ($L = F(f\,; U = U_i, T = 0)$)

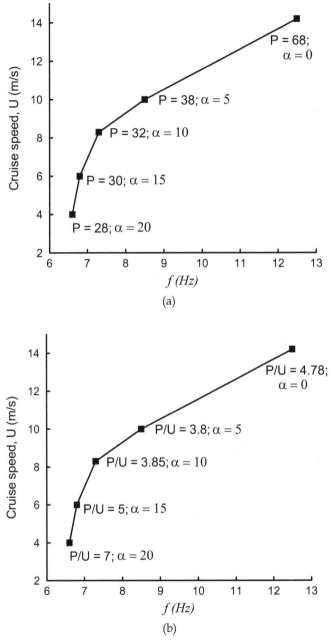

Fig. 11. Cruise speed versus flapping frequency for $L = W = 0.15 \, kg$ ($T = 0$), (a): value of P and α at each point, (b): value of $\dfrac{P}{U}$ and α at each point

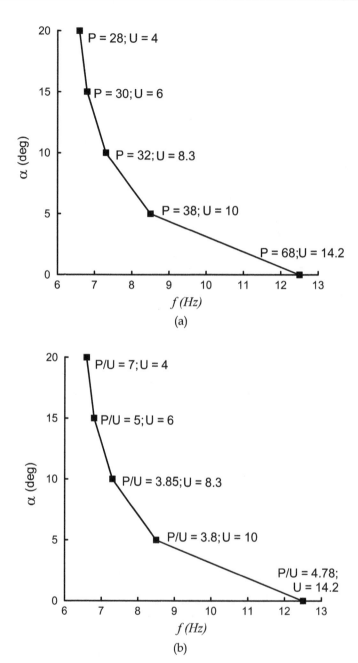

Fig. 12. Angle of incidence versus flapping frequency for $L = W = 0.15\ kg$ ($T = 0$), (a): value of P and U at each point, (b): value of $\dfrac{P}{U}$ and U at each point

For a given vehicle weight ($W = 0.15\ kg$), these diagrams are shown in Figs. 13 and 14. As Fig. 13 shows, for the highest endurance, assuming maximum allowable angle of incidence

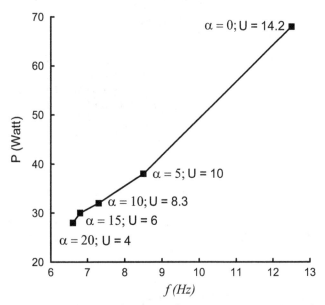

Fig. 13. Power usage versus flapping frequency for $L = W = 0.15\ kg\ \ (T = 0)$

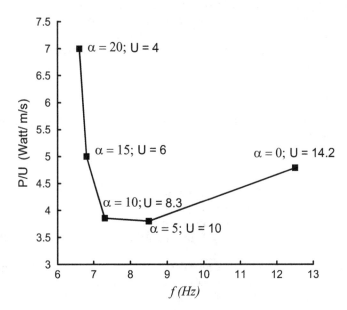

Fig. 14. $\dfrac{P}{U}$ versus flapping frequency for $L = W = 0.15\ kg\ \ (T = 0)$

equal to 20°, this vehicle may fly by 4.0 m/s cruise speed at 20° angle of incidence. Fig. 14 also shows that for maximum range (minimum $\frac{P}{U}$), the vehicle should fly at 8.3 m/s with 10° angle of incidence. Of course these results are correct only for our example wing, but the scheme used here is universal.

As a summary, one may consider two criteria for optimum flight conditions, i.e. flight endurance and range. For each criterion, one may find two flight curves. These curves show optimum flight conditions. Here, we have found only one point of these curves which corresponds to W equal to 0.15 kg. Finding the other points is an easy task, which is not needed here.

6. Conclusion

A flapping wing system and an experimental set-up are designed to measure the lift, thrust and power usage of the flapping wing motion for different flapping frequencies, angles of attack and various wind tunnel velocities up to 12 m/s. Several new parameters are introduced for flapping wings and different criteria are presented for performance analysis. Flight envelopes for a particular flapping vehicle and a new scheme for finding them is introduced. An algorithm is proposed so that for each given mission and vehicle weight (or payload), one may find the best cruise speed, angle of incidence or flapping frequency for maximum range, which is corresponding to minimum ratio of power to cruise speed (i.e,. energy usage for unit distance), or for maximum endurance corresponding to minimum power (i.e., minimum flapping frequency). This scheme allows us to show cruise speed as a function of the flapping frequency, for various angles of incidence, and energy usage over unit distance as a function of the flapping frequency for various angles of incidence. For a special mission with 0.15 kg weight, the suitable cruise speed versus flapping frequency is demonstrated. The resulting diagrams are used to find optimum performance of the flapping wing vehicle.

7. Acknowledgment

Authors acknowledge the support of the Sharif University of Technology and Aerospace Engineering Department.

8. References

Aditya, K. & Malolan, V. (2007). Investigation of Strouhal Number Effect on Flapping Wing Micro Air Vehicle. *AIAA Paper 2007-486*

Ames, R. (2001). On the Flowfield and Forces Generated by a Rectangular Wing Undergoing Moderate Reduced Frequency Flapping at Low Reynolds Number. Ph.D. Thesis, Georgia Institute of Technology

Daniel, T. & Combes, S. (2002). Flexing Wings And Fins: Bending By Inertial or Fluid Dynamic Forces ? *Integrative and Comparative Biology* , 42, 1044-1049

Fritz, T. E. & Long, L. N. (2004). Object-Oriented Unsteady Vortex Lattice Method for Flapping Flight. *Journal of Aircraft*, 41 (6), 1275–1290

Gallivan, P. & DeLaurier, J. (2007). An Experimental Study of Flapping Membrane Wings. *Canadian Aeronautics and Space Journal*, 53 (2), 35-46

Gebert, G. & Gallmeier, P. (2002). Equation of Motion for Flapping Flight. *AIAA Paper 2002-4872*

Ho, S. (2003). Unsteady Aerodynamics and Adaptive Flow Control of Micro Air Vehicles. Ph.D. Thesis, University of California

Hong, Y. S. & Altman, A. (2007). Streamwise Vorticity in Simple Mechanical Flapping Wings. *Journal of Aircraft*, 44 (5), 1588-1597

Isaac, K. Colozza, A. & Rolwes, J. (2006). Force Measurements on a Flapping and Pitching Wing at Low Reynolds Numbers. *AIAA Paper 2006-450*

Isogai, K. & Harino, Y. (2007). Optimum Aeorelastic Design of a Flapping Wing. *Journal of Aircraft*, 44 (6), 2040-2048

Jones, K. D., Castro, B. M., Mahmoud, O., Pollard, S. J., Platzer, M. F., Gonet, K., et al. (2002). A Collaborative Numerical and Experimental Investigation of Flapping-Wing Propulsion. *AIAA Paper 2002-0706*

Kim, D. K., Kim, H. I., Han, J. H. & Kwon, K. J. (2008). Experimental Investigation on the Aerodynamic Characteristics of a Bio-mimetic Flapping Wing with Macro-Fiber Composites. *Journal of Intelligent Material Systems and Structures*, 19 (3), 423-431

Kim, S., Jang, L., Kim, M. & Kim, J. (2003). Patent No. US 6,550,716 B1

Lasek, M. & Sibliski, K. (2003). Analysis of Flight Dynamics and Control of an Entomopter. *AIAA Paper 2003-5707*

Lin, C. S., Hwu, C. & Young, W. B. (2006). The Thrust and Lift of an Ornithopter's Membrane Wings with Simple Flapping Motion. *Aerospace Science and Technology*, 10, 111-119

Mantia, M. L. & Dabnichki, P. (2009). Unsteady Panel Method for Flapping Foil. *Engineering Analysis with Boundary Elements*, 33, 572-580

Mazaheri, K. & Ebrahimi, A. (2010a). Experimental Investigation of the Effect of Chordwise Flexibility on the Aerodynamics of Flapping Wings in Hovering Flight. *Journal of Fluids and Structures*, 26, 544-558

Mazaheri, K. & Ebrahimi, A. (2010b). Experimental Study on Interaction of Aerodynamics With Flexible Wings of Flapping Vehicles in Hovering and Cruise Flight. *Archive of Applied Mechanics*, 80, 1255-1269

Muniappan, A., Baskar, V. & Duriyanandhan, V. (2005). Lift and Thrust Characteristics of Flapping Wing Micro Air Vehicle (MAV). *AIAA paper 2005-1055*

Murphy, J. (2008). Experimental Investigation of Biomimetic Wing Configurations for Micro Air Vehicle Applications. Master of Science Thesis, Aerospace Engineering, Iowa State University

Pennycuick, C. J. (2008). *Modelling the Flying Bird*. Academic Press

Pfeiffer, A. T., Lee, J. S., Han, J. H. & Baier, H. (2010). Ornithopter Flight Simulation Based on Flexible Multi-Body Dynamics. *Journal of Bionic Engineering*, 7 (1), 102-111

Shyy, W., Berg, M. & Ljungqvist, D. (1999). Flapping and Flexible Wings for Biological and Micro Air Vehicles. *Progress in Aerospace Science*, 35, 455-505

Shyy, W., Lian, Y., Tang, J., Viieru, D. & Liu, H. (2008). *Aerodynamics of Low Reynolds Number Flyers*. Cambridge University Press

Singh, B. (2006). Dynamics and Aeroelasticity of Hover-Capable Flapping Wings: Experiments and Analysis. Ph.D. Thesis, University of Maryland

Smith, M. J., Wilkin, P. J. & Williams, M. H. (1996). The Advantages of an Unsteady Panel Method in Modelling the Aerodynamic Forces on Rigid Flapping Wings. *Journal of Experimental Biology*, 199, 1073–1083

Vest, M. S. & Katz, J. (1996). Unsteady Aerodynamic Model of Flapping Wings. *AIAA Journal*, 34 (7), 1435-1440

Willis, D. J., Peraire, J., Drela, M. & White, J. K. (2006). A Numerical Exploration of Parameter Dependence in Power Optimal Flapping Flight. *AIAA Paper 2006-2994*

Wilson, N. L. & Wereley, N. (2007). Experimental Investigation of Flapping Wing Performance in Hover. *AIAA Paper 2007-1761*

Wu, P., Stanford, B. & Ifju, P. (2008). Structural Deformation Measurements of Anisotropic Flexible Flapping Wings for Micro Air Vehicles. *AIAA Paper 2008-1813*

Permissions

The contributors of this book come from diverse backgrounds, making this book a truly international effort. This book will bring forth new frontiers with its revolutionizing research information and detailed analysis of the nascent developments around the world.

We would like to thank Dr. Jorge Colman Lerner and Dr. Ulfilas Boldes, for lending their expertise to make the book truly unique. They have played a crucial role in the development of this book. Without their invaluable contribution this book wouldn't have been possible. They have made vital efforts to compile up to date information on the varied aspects of this subject to make this book a valuable addition to the collection of many professionals and students.

This book was conceptualized with the vision of imparting up-to-date information and advanced data in this field. To ensure the same, a matchless editorial board was set up. Every individual on the board went through rigorous rounds of assessment to prove their worth. After which they invested a large part of their time researching and compiling the most relevant data for our readers. Conferences and sessions were held from time to time between the editorial board and the contributing authors to present the data in the most comprehensible form. The editorial team has worked tirelessly to provide valuable and valid information to help people across the globe.

Every chapter published in this book has been scrutinized by our experts. Their significance has been extensively debated. The topics covered herein carry significant findings which will fuel the growth of the discipline. They may even be implemented as practical applications or may be referred to as a beginning point for another development. Chapters in this book were first published by InTech; hereby published with permission under the Creative Commons Attribution License or equivalent.

The editorial board has been involved in producing this book since its inception. They have spent rigorous hours researching and exploring the diverse topics which have resulted in the successful publishing of this book. They have passed on their knowledge of decades through this book. To expedite this challenging task, the publisher supported the team at every step. A small team of assistant editors was also appointed to further simplify the editing procedure and attain best results for the readers.

Our editorial team has been hand-picked from every corner of the world. Their multi-ethnicity adds dynamic inputs to the discussions which result in innovative outcomes. These outcomes are then further discussed with the researchers and contributors who give their valuable feedback and opinion regarding the same. The feedback is then collaborated with the researches and they are edited in a comprehensive manner to aid the understanding of the subject.

Apart from the editorial board, the designing team has also invested a significant amount of their time in understanding the subject and creating the most relevant covers. They scrutinized every image to scout for the most suitable representation of the subject and create an appropriate cover for the book.

The publishing team has been involved in this book since its early stages. They were actively engaged in every process, be it collecting the data, connecting with the contributors or procuring relevant information. The team has been an ardent support to the editorial, designing and production team. Their endless efforts to recruit the best for this project, has resulted in the accomplishment of this book. They are a veteran in the field of academics and their pool of knowledge is as vast as their experience in printing. Their expertise and guidance has proved useful at every step. Their uncompromising quality standards have made this book an exceptional effort. Their encouragement from time to time has been an inspiration for everyone.

The publisher and the editorial board hope that this book will prove to be a valuable piece of knowledge for researchers, students, practitioners and scholars across the globe.

List of Contributors

Harun Chowdhury
School of Aerospace, Mechanical and Manufacturing Engineering, RMIT University, Australia

José Ignacio Íñiguez, Ana Íñiguez-de-la-Torre and Ignacio Íñiguez-de-la-Torre
Universidad de Salamanca, Departamento de Física Aplicada, Spain

Efstathios Konstantinidis and Demetri Bouris
Department of Mechanical Engineering, University of Western Macedonia, Bakola & Sialvera, Kozani, Greece

Maxime Huet, Gilles Rahier and Francois Vuillot
Onera – The French Aerospace Lab, F-92322 Châtillon, France

M. Casper and P. Scholz
Institute of Fluid Mechanics, Braunschweig Technical University, Germany

J. Colman, J. Marañón Di Leo, S. Delnero and M. Camocardi
Boundary Layer and Environmental Fluid Dynamics Laboratory, Engineering Faculty, National University of La Plata, Argentina

Sofiane Khelladi, Christophe Sarraf, Farid Bakir and Robert Rey
DynFluid Lab., Arts et Métiers ParisTech, France

J. Lassig
Environmental Fluid Dynamics Laboratory, Engineering Faculty, National University of Comahue, Argentina

J. Colman
Boundary Layer and Environmental Fluid Dynamics Laboratory, Engineering Faculty, National University of La Plata, Argentina

Oscar Falcinelli and José Tamagno
Aeronautical Department, FCEFyN, Nacional University of Cordoba, Argentina

Sergio Elaskar
CONICET, Argentina Aeronautical Department, FCEFyN, Nacional University of Cordoba, Argentina

Jorge Colman Lerner
Aeronautical Department, Engineering Faculty, Nacional University of La Plata, Argentina

Abbas Ebrahimi and Karim Mazaheri
Aerospace Systems Center of Excellence, Sharif University of Technology, Iran

Printed in the USA
CPSIA information can be obtained
at www.ICGtesting.com
JSHW011401221024
72173JS00003B/369